Digital Signatures

Digital Signatures

The Impact of Digitization on Popular Music Sound

Ragnhild Brøvig-Hanssen and Anne Danielsen

The MIT Press
Cambridge, Massachusetts
London, England

© 2016 Massachusetts Institute of Technology

All rights reserved. No part of this book may be reproduced in any form by any electronic or mechanical means (including photocopying, recording, or information storage and retrieval) without permission in writing from the publisher.

This book was set in Stone by the MIT Press.

Library of Congress Cataloging-in-Publication Data

Names: Brøvig-Hanssen, Ragnhild. | Danielsen, Anne.
Title: Digital signatures : the impact of digitization on popular music sound / Ragnhild Brøvig-Hanssen and Anne Danielsen.
Description: Cambridge, MA : The MIT Press, 2016. | Includes bibliographical references and index.
Identifiers: LCCN 2015038376 | ISBN 9780262034142 (hardcover : alk. paper) ISBN 9780262549639 (paperback)
Subjects: LCSH: Popular music—Philosophy and aesthetics. | Popular music—Production and direction. | Sound—Recording and reproducing—Digital techniques.
Classification: LCC ML3877 .B77 2016 | DDC 781.640285—dc23 LC record available at http://lccn.loc.gov/2015038376

Contents

Preface and Acknowledgments   vii

1 Introduction: Digital Technology and Popular Music Sound   1

2 Making Sense of Digital Spatiality: Kate Bush's Eerie Collage   21

3 The Instrument Formerly Known as the Machine: Hyperaccuracy and Sonic Richness in Prince's "Kiss"   43

4 The Rebirth of Silence in the Company of Noise: Portishead Going Retro   61

5 Cut-Ups and Glitches: The Freeze and Flow of Los Sampler's and Squarepusher   81

6 Seasick Computers: Microrhythmic Manipulation in the Era of Endless Undo   101

7 Autotuned Voices: Alienation and "Brokenhearted Androids"   117

8 Popular Music in the Digital Era   133

Notes   153
References   169
Index   183

Preface and Acknowledgments

Music is always immersed in technology. Both authors of this book, for a very long time, have been fascinated by the various musical manifestations of this fact. The presence and mediation of musical technology are sometimes concealed by the music itself, and/or by its discourse, and at other times exposed, even to the degree that the musical technology occupies the listener's focus of attention. In either case, technology contributes significantly to the musical expression. Through a close reading of examples from the evolving practice of digital music production over the last thirty or forty years, we aim to illuminate the ways in which the newly developed techniques of digital sound technology have enabled a transformation of popular music sound. The book is addressed to a crossover audience of scholars and students across a range of academic disciplines (including popular music studies, music technology and record production, sound studies, and media and cultural studies), to media professionals and music practitioners, and to the general nonspecialist audience. We hope that this book will not only contribute to scholarship and research in the field but also encourage more of it.

This book started as an outgrowth of Ragnhild Brøvig-Hanssen's PhD thesis, "Music in Bits and Bits of Music: Signatures of Digital Mediation in Popular Music Recordings" (2013), for which Anne Danielsen served as supervisor. Throughout the thesis-writing process, we had many fruitful discussions related to our common interest in the various ways in which technology contributes to sound, and thus to our musical experience in general, and the impact of digitization on popular music sound in particular. After the completion and acceptance of the thesis, we both saw a need to extend the material and widen its potential outreach. Three seminal chapters from Brøvig-Hanssen's thesis (chapters 2, 4, and 5 in the present volume) were revised and complemented by three chapters written by Anne Danielsen (chapters 3, 6, and 7). Newly revised introduction and conclusion chapters were provided by Brøvig-Hanssen, aided and supplemented by Danielsen, and then the entire contents and direction of the manuscript were adjusted, rewritten, and molded into the present book.

Many individuals made our work both possible and pleasurable. First and foremost, we wish to thank Douglas Sery of the MIT Press for his initial belief in this book and continually positive approach. We also want to thank the press's anonymous reviewers for providing valuable feedback. We are very grateful to Nils Nadeau, who copyedited the manuscript; in addition to his thorough and well-considered enhancement of its prose, he offered clever and constructive queries that enabled us to improve its content considerably. Our thanks also go to Hildegunn Lauve Hansen and Peter Knudsen for designing and helping with the illustrations. We found much inspiration at several Art of Record Production (ARP) conferences and want to thank the audiences of those conferences for their feedback on our papers, and for presenting inspiring papers in turn. We also want to thank the ARP conference organizers, Katia Isakoff and Simon Zagorski-Thomas, as well as the ASARP association for their important initiatives in supporting, encouraging, and networking the emerging field of the study of record production. Finally, we are very grateful to all of our colleagues at the Department of Musicology at the University of Oslo, in both the academic and the administrative staff, for the pleasure of their company, for the many interesting and inspiring (academic and nonacademic) discussions, and for their help and valuable insight. We also want to thank our students, from whom we have learned much.

In addition, Ragnhild Brøvig-Hanssen wants to thank the Faculty of Humanities for providing her with the financial support to embark on the stimulating three-year journey that produced the PhD thesis upon which this book is based. In addition to Anne Danielsen, she is very grateful to Simon Frith, Bernard Gendron, Lydia Goehr, Stan Hawkins, Hedda Høgåsen-Hallesby, Tellef Kvifte, and Susan McClary for their sharp-eyed readings and valuable feedback on early drafts of the initial thesis. She would also like to thank the appointed committee members of the PhD thesis—Nicola Dibbon, Stan Hawkins, and Albin Zak—for their constructive and detailed feedback, which proved to be particularly useful in the further elaboration of that manuscript into the present book. Finally, she wants to express her profound appreciation to her family and friends for their steadfast and unflagging support. Very special thanks go to her husband, Henning, and to her two daughters, Hedvig and Agnes.

Anne Danielsen is very grateful to Gisela Attinger, Gary Bromham, Richard J. Burgess, Erling Guldbrandsen, Stan Hawkins, Jørgen Langdalen, Preben von der Lippe, and Tellef Øgrim for their valuable comments on different parts of the manuscript. She also wants to thank Christer Falck and Petter Aagaard for stimulating discussions about Prince's music, and Mari Paus for her contribution to the initial phase of background research for chapters 3, 6, and 7. She also thanks her family, above all, for their patience and encouragement.

RBH and AD
Oslo, April 2015

1 Introduction: Digital Technology and Popular Music Sound

Throughout popular music's history, technological mediation has been imperative to its production, as well as its distribution and consumption. New developments in technology have always had an immediate impact on the art form, and digital technology is, of course, no exception. Obviously, the development of affordable computer-based digital audio workstations (DAWs)—a software recording workstation running on computers with the possibility of audio and MIDI interface hardware—together with the development of the Internet, has had consequences for where, when, and by whom music is made, as well as how it is distributed and acquired. Consequently, the control and power of the professionals and their studio structures have been decentralized by an amorphous figure whom Paul Théberge labels the "singer-songwriter-producer-engineer-musician-sound designer" (Théberge 1997, 221–222). The stereotypical narrative of the digital age of music and recording practices is that the "amateur bedroom producer" downloads preexisting music from the Internet via a peer-to-peer network, deconstructs and remixes the music on her or his laptop, and distributes and promotes the new version through Internet file-sharing sites from which fans download it, escaping industry control and undermining copyright laws.[1] Equally important, however, are the ways in which this development has affected how the music *sounds*. This book sets out to present new insights into the impact of the digitization of technology on the *aesthetics* of popular music.

The digital recording medium converts sound signals into binary numbers, then reconverts the numbers back into sound signals (we will return to this conversion process shortly). This conversion of sounds into numbers has clearly affected how music is produced, because digitally converted sounds can be treated differently from analog sounds. Digital technology has, for instance, provided more opportunities to control combinations of sounds, and to control every aspect of an individual sound, which has both made it easier to conceal any traces of manipulation and introduced new possibilities for manipulating sounds in audible and experimental ways. In this book we address the musical moments when the use of digital technology is *revealed* to the listener. We refer to these musical aspects that bear clear, audible traces of digital technology as *digital signatures*—that is, the sonically distinctive character of digital mediation. Put simply, digital signatures are the sonic fingerprints of digital technology.

The particular signatures of digital mediation upon which we will focus include digital reverb and delay, MIDI and sampling, digital silence, the virtual cut-and-paste tool, digital glitches, microrhythmic manipulation, and autotuning. To this end, we will examine these digital signatures at work in particular musical texts. We have singled out selected songs by artists that all belong to, or at least reside on the outskirts of, the Anglo-American mainstream, including Kate Bush, Prince, Portishead, Los Sampler's (Uwe Schmidt), Squarepusher, Snoop Dogg, Bon Iver, and Lady Gaga. The tracks are chosen first and foremost because they illustrate and highlight the different forms of digital signatures that we explore. In other words, we have privileged music in which a given digital technology has been both experimented with and exposed. A general trend is that when a technological device is new, it is often subjected to radical experimentation, despite the fact that it was often developed to solve a problem in a transparent manner, and our selected tracks are therefore often from the same period that the technological tool in question was introduced. However, some of the digital signatures that we discuss represent instead a revival of an effect introduced several years earlier. In addition, then, to representing different digital signatures, the selected tracks also represent important stages of the development of digital music devices and practices from the early 1980s to the present.

In tandem with our discussion of the technical aspects of the selected digital tools and techniques, as well as the ways in which they have been used in particular musical texts, we also seek to explore how the different forms of digital mediation manifest the potential to generate completely new musical meanings. In particular, we will focus on three strands of musical meaning that have been afforded by the ways in which the tools, techniques, and sounds typical of the digital era have been used. The first strand concerns how the digitization of technology has supplied a new musical language or compositional palette. An example of this is the unique degree and quality of expressivity that came about as a consequence of new opportunities for editing rhythmic events at the level of microseconds with DAWs, resulting in both the "seasick" rhythms that were prevalent in African-American musical styles such as hip hop, neo-soul, and contemporary R&B toward the end of the millennium and the various cut-and-paste rhythms that are now common in electronica music styles such as IDM (so-called intelligent dance music) and glitch. The second strand concerns the fact that the digital era has generated a renewed sense of space and time by emphasizing, through digital technology's more convenient means of experimentation, the sounds' liberation from the spatiotemporal coherence through which they were once constrained. In a sense, the digitization of technology has *materialized* musical space and time via numerical information that can be mathematically manipulated, rearranged, and juxtaposed in this digitally nondestructive, editable environment. The third strand concerns the fact that digital technology has once again managed to challenge the discursive dichotomy of human versus machine. While the morphing and blurring of these categories has been a topical theme in popular music studies for decades, experimentation with digital signatures presents an opportunity to revisit these discussions from a new and updated perspective. Discussions of these three strands of musical meaning will recur in various forms and through various perspectives in the music analyses of this book, then return in force in the concluding chapter.

In classical music, folk, and jazz, the recording medium has traditionally had a documentary function, in that its main purpose has been to capture a given live performance.[2] Popular music, on the other hand, has been virtually determined by technological mediation; the recording is, as Allan F. Moore and Theodore Gracyk among others have pointed out, the music's "primary text" (Moore 2001, 34–35) or "primary medium" (Gracyk 1996, 21). In the present study, we are concerned with the technological mediation that happens within the production process of music—that is, the processes of recording, editing, and treating sounds with various signal-processing effects. Sometimes the sonic imprints of technological mediation are obvious to the listener; other times they affect the sound without being perceived as such. In either case, though, mediating technology is hugely important to musical expression. Despite the obvious impact that technological mediation has had on sounds, technology has not been given the attention it deserves in music analysis. Whereas several scholars have contributed significantly to the conversation regarding how technological mediation has shaped popular music practices over the last twenty years, this domain has generally been approached from a sociological or historical orientation with a focus on extramusical features rather than the music itself. For instance, a central issue in the scholarly discussion has been how changes in recording technologies have had industrial and political consequences that have led in turn to changes in the industry's organization and consequently its power structures. Such studies have concerned themselves with the ways in which digital technologies have influenced how music is marketed, promoted, distributed, and acquired, and how technology has affected the economic structure of the music industry in terms of skirting the traditional path that leads to copyright royalties.[3] Technological innovation, in terms of recording technologies and musical instruments, has also interested many scholars, especially with regard to the ways in which sound technologies are socially and economically embedded.[4] Another central question involving the consumption practices of music has been how new technologies have altered listening habits and the listener's approach to music as reified, and consequently how the boundary between production and consumption has started to blur.[5] While several scholars, particularly those with a background in music criticism, ethnomusicology, history, or cultural studies, have given impressive accounts of music's cultural contexts and social and economical effects, they have been hard pressed to relate such extramusical features to intramusical features. That is, the *musical* effects of technological mediation generally escape the focus of these studies.

In line with this, methodological approaches such as aesthetic interpretation and music analysis are seldom introduced into discussions of music technology. Conversely, few scholars interested in the former pay adequate attention to the technological mediation involved in the music they examine; in retrospect, it is truly remarkable how little consideration has been given to technological aspects in the field of music analysis. This trend is starting to change—we now see an emerging research field involving the art of record production—but huge gaps remain in the literature, and we continue to lack adequate methodological approaches and, in particular, a conceptual framework for analyzing the technological aspects of popular music production.

In this book, we will present and discuss the opportunities and effects of the new technological means on popular music sound through music analyses conducted within the academic tradition of subjective interpretive hermeneutics. We have chosen this methodological approach because it presents a unique opportunity to generate a nuanced and detailed account of the musical texts at hand, as well as to grasp the (sometimes) subtle stylistic and cultural meanings that emerge from these texts. Through theorizing the role of technological mediation in music recordings, we will attempt to elucidate how technological mediation in general, and digital mediation in particular, contributes to the aesthetics of popular music.

Digital Signatures and Opaque Mediation

As mentioned, digital signatures encompass the sonically distinctive character of digital mediation. Digital signatures are thus an example of what Ragnhild Brøvig-Hanssen elsewhere has called "opaque mediation" (2010, 2013a, forthcoming). Building on the French philosopher Louis Marin,[6] she has distinguished between opaque and transparent mediation to clarify that what is usually at stake is not whether the music is technologically unmediated or mediated, or how much technological mediation is involved, but rather how the technological mediation in the music is *experienced*. While "opaque" and "transparent" mediation are experiential categories, they also signal divergent aesthetic paradigms at work in the production process of music. Marin illustrates the conditions of transparency and opacity with the metaphor of a pane of glass: a clean windowpane through which we look at the landscape beyond is both present and absent—it is there, but it is not our focus of attention; we perceive it as transparent. If there are scratches or stains or blotches on it, however, our attention will be diverted to the pane itself, rather than the landscape outside (Marin 1991, 57). Likewise, if one's production ideal is transparent mediation, then technology will be used in a manner that allows the listener to ignore it. If one's production ideal is opaque mediation, then technology will be used in a manner that forces the listener to reckon with it. Here the aesthetic potential of the technological mediation's self-presentation is dedicated to the production of unique musical effects—the technological mediation has a voice of its own, in fact, and insists on its role in the experiential meaning of the music.

The *experiential* aspect of the concept of a "digital signature" of mediation is crucial and must not be confused with a "digital feature" of mediation. Whereas a digital feature basically indicates that the mediation involved in the music originates in processes that involve digital technology, a digital signature points instead to a musical sound that is experienced as a footprint of the digital. That is, a digital signature is an act of technological mediation experienced as opaque and associated with the digital. This means that a digital feature need not be experienced as a digital signature; it might also appear to be transparent. Moreover, an act of technological mediation that is experienced as a digital signature does not necessarily mean that it belongs exclusively to the digital domain; whereas some digital signatures are unique to the digital medium, others have resulted from the ways in which digital technology has reinvented analog tools and techniques or made them more common. For instance, while the cut-and-paste technique is not unique to digital technology, the scale with which this technique is applied within the digital era is almost unthinkable via analog technology, and it is thus often associated with the digital. The resulting *quantitative* change in its use is so dramatic that it has, in a sense, become a *qualitative* signature of the digital.

As mentioned, the digital signatures that we have chosen to focus on in this book include digital reverb and delay, MIDI and sampling, the characteristic digital silence, the virtual cut-and-paste tool and digital glitches, microrhythmic manipulation, and autotuning. Of course, whether these sonic effects are heard as the result of digital mediation depends on the listener. Our aim, however, is to explain some of the ways in which the digitization of technology has affected popular music sound in general, and some of the reasons why the sonic effects resulting from the exposure of these tools or techniques are often heard as digital signatures. Accordingly, in addition to explaining the characteristic sound of each of the digital signatures presented, we will situate them in a historical perspective by discussing how their functions and constraints differ from predigital music devices.

Even if the listener does not experience these signatures as digital, he or she is likely to notice them—that is, to hear them as instances of opaque mediation. Yet this aspect of mediation may also vary with the listener and the context. Some of us are likely to focus on particular forms of technological mediation rather than others, whereas others might ignore those same forms, and all of this could change over periods of time that range from minutes to decades or more. For instance, when the microphone was introduced in the mid-1920s, singers were quick to exploit the microphone's capacity for mediation as a personal instrument of sorts, developing new vocal styles such as "crooning," which emphasizes the uniquely public intimacy of the amplified voice. While this use of the voice simply mimics in a musical context the way people hear the voice in unmusical contexts (face-to-face conversations, for example), it was at first regarded as a profoundly opaque mediation, since the intimate voice had never before been able to penetrate throughout a concert hall. This familiar-made-unfamiliar vocal sound did not correspond to the singer's spatial location, either; though the vocalist sang from a stage in a concert hall, far away from the listener, it sounded as if he or she were sitting right next to the listener. As listeners grew accustomed to such live performances, the microphone-staged voice gradually became naturalized[7] and came to stand for the musical voice itself.[8] Therefore, the technological mediation grew increasingly transparent, in the sense that, perceptually, it passed by the listener unnoticed.

Factors other than the historical also determine whether we perceive a particular act of technological mediation as transparent or opaque; standards vary across genres as well. Fans of acoustic jazz may perceive the aggressive use of the compressor as opaque, whereas fans of contemporary pop music may perceive it as transparent. Nevertheless, although opacity and transparency are obviously not inherent qualities of music, one's comprehension of technological mediation as either opaque or transparent is far from arbitrary, as we will demonstrate throughout this book. In fact, we usually experience technological mediation as opaque at those moments when it disturbs our mental imagination of the sound source's "pure" identity (that is, when it crosses the border between what we experience as "sound" and as "mediation"); when it challenges our notions of "extramusical mediation" (mediation that is not part of the music) and "intramusical mediation" (mediation that belongs to the musical production); and when it disrupts the spatiotemporal coherence of the music. In the end, whether the technological mediation draws attention to itself depends on the perceiver, the music, and contextual factors.

"Opaque" and "transparent" mediation might evoke Denis Smalley's distinction between the "naturalist work" and the "interventionist work":

"At one extreme, a naturalist work unfolds as if 'natural,' with few seams and ruptures, and a logic of passage; there is a certain transparency in the way things proceed, above all in the care with mixing. With the interventionist approach the composer's hand is in evidence, and the stamp of the technology and techniques is apparent in the kind of material and the way it is manipulated, whereas in the naturalist work there will be some attempt to hide techniques, and avoid exposing technological signifiers" (Smalley 2007, 54). Although Smalley's "naturalist work" sounds like an example of transparent mediation, and his "interventionist work" sounds quite like opaque mediation, his terms are problematic. First of all, the means of achieving transparent mediation might involve just as much "intervention" as those achieving opaque mediation. Furthermore, in relation to the connotations of the "naturalist work," we must recognize that opaque mediation is experienced as both unnatural and natural, depending on all of the factors listed above, and therefore, as a qualifier, "natural" has little to recommend it. The notion of "opaque mediation" might further evoke Mark Katz's notion of "phonograph effects"—that is, "the manifestations of sound recording's influence," which he introduces in *Capturing Sound: How Technology Has Changed Music* (2004) in his attempt to illuminate how and why production influences music and listeners (Katz 2004, 3). In some of his case studies, such as his analysis of "Praise You" by Fatboy Slim, the term "phonograph effect" could in fact be replaced by "opaque mediation" or even "digital signature," and the same holds true for several of our analyses as well. Yet an important distinction remains: "phonograph effect" describes *any* influence that technology has had on music and the listener, such as how the three-minute limit of a ten-inch 78-rpm phonograph record dictated (and, following Katz, still influences) the length of the popular song (ibid., 32), or how the MP3 format and P2P network have provided new ways of both disseminating and experiencing music (ibid., 158–187). Opaque mediation, on the other hand, only describes the technological mediation involved in the musical production that is experienced as exposed. Simply put, all instances of opaque mediation are phonograph effects, but all phonograph effects are not instances of opaque mediation. In parallel, all digital signatures are opaque mediation, but all instances of opaque mediation are not instances of a digital signature since opaque mediation encompasses all kinds of technologies used in the production process that are exposed, analog as well as digital.

Digital Technology Enters the Mainstream

Digital technology found its way to music devices already in the 1960s, when American scientist Thomas Stockham began to experiment with digital audio recording systems. When talking about the "digital revolution" in the field of music, however, we refer not to this initial phase of digitization but to the cultural turn that gradually arose in the 1980s and the 1990s and encompasses digital synthesizers, digital drum machines, digital sampler instruments, the MIDI system, digital signal processing (DSP) effects, the CD and MP3 formats, the computer- and software-based recording platform often referred to as the digital audio workstation (DAW), and the new musical arenas introduced by the Internet (YouTube, Myspace, P2P networks, and so on).

Digital recording is based on the pulse-code modulation (PCM) of a sound signal, which was devised by Alec Reeves in 1939 (Watkinson 1999, 112). Unlike both the phonograph and the magnetic tape recorder, which store continuous sound signals, the digital recording medium converts sound signals into numbers in the binary number system—that is, streams or bytes consisting of combinations of zeroes and ones. This process requires a circuit that converts the analog sound into digital samples, a so-called analog-to-digital converter (ADC), where the continuous sound signal is sampled and stored as digital information (see figure 1.1). This means that the voltage of the sound's waveform, which is basically the continuous rise and fall of amplitudes over time, is measured (or "sampled") at several exactly uniform time increments. The rate at which the waveform is sampled (the amount of samples per second) determines the sound quality of the recording, because it affects the frequency response. With modern digital recording technology, the sound signal is sampled 44,100 times a second (44.1 kHz), which results in a rather accurate reconstruction of the original sound signal.[9] In addition to the sampling rate, the number of intervals (the bit depth) used to measure (or "quantize") the dynamic range determines the accuracy of each sample. As mentioned above, each measurement is represented by a number in the binary number system—that is, as a string (byte) of binary digits ("bits"). The higher the bit depth, the more intervals there are in the scale, and the better the sound quality. With the earliest samplers, producers could not sample with more than 8-bit accuracy, which represents 256 intervals (2^8) and a relatively low sound quality.

The standard in the audio industry today is to sample with 24-bit accuracy, which represents 16,777,216 intervals (2^{24}), when recording sounds, and to sample with 16-bit accuracy, which represents 65,536 intervals (2^{16}), when bouncing the sounds to the CD format (Pohlmann 2000, 32–36). When the recorded material is played back, a digital-to-analog converter (DAC) reconverts the digital information to continuous sound waves, re-creating the exact voltage levels of the sound wave by calculating the sampled and quantized information.

In the beginning, the digitally converted sounds were stored on tape,[10] but it soon became possible to store the information on floppy disks and hard disks and in microprocessors as well. The first practical digital recorder was demonstrated in 1967 (Watkinson 1999, 122), but, because of its great expense, it did not become standard technology in recording studios until the early 1990s (Millard 2005, 356). However, the digital sampler instrument, digital synthesizers and drum machines, and digital signal processing (DSP) effects were embraced by the music industry already in the late 1970s and 1980s.

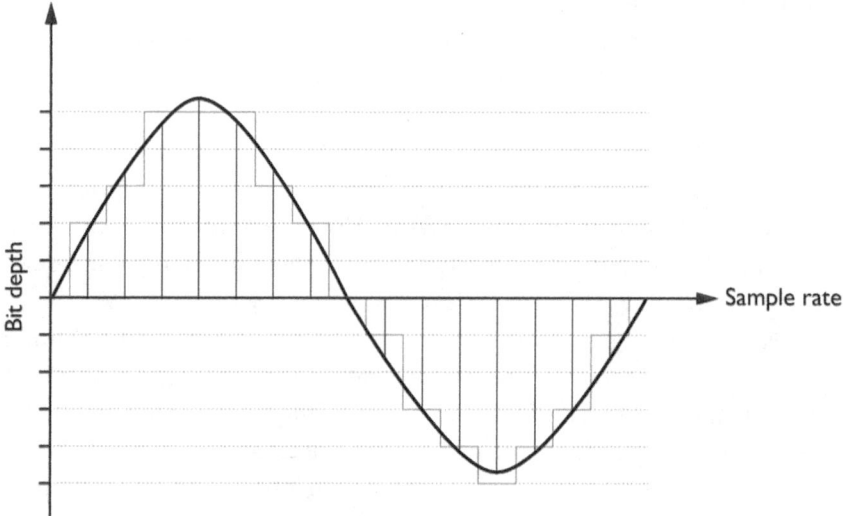

Figure 1.1
When a sound is converted from the analog to the digital domain, the amplitudes of the sound wave are sampled according to a given frequency (the sampling rate) and measured according to a scale consisting of a given number of intervals of the dynamic range (the bit depth). The vertical lines here represent the sampling rate, while the horizontal lines represent the bit depth.

The sampler instrument, which is essentially a digital recording medium, was introduced at the end of the 1960s, but initially it could sample only very short sound sequences, owing to its extremely limited random-access memory; for example, the Fairlight Computer Musical Instrument (CMI), first available in 1979, allowed only for a second or two of sound. Early in the 1980s, then, music makers and manufacturers of musical instruments exploited the sampling technique mainly in terms of sampling short sounds from instruments in order to replace the existing synthetic sounds with more realistic ones (Rose 1994, 73; for a more detailed account of early digital sampling, see chap. 3). Digital drum computers (such as the Linn LM-1 or Oberheim DMX, introduced in the early 1980s) and digital synthesizers (such as the Kurzweil or the Yamaha DX7, introduced in 1983) soon followed; these were based on samples (digitally recorded sounds) and frequency modulation (FM) synthesis. It was, in fact, not until the E-mu SP-12 became available in 1986 that one could sample even up to ten seconds of sound (Fernando 1994, 225). By the late 1980s, as the storage capacity of samplers had increased and the price had decreased, the sampler instrument had become a popular asset among musicians. The sampler instrument introduced us to experiment with a new means of constructing music out of bits and pieces, a practice that was also encouraged by the digital audio workstation.

In 1981, the companies Sequential Circuits, Oberheim, and Roland agreed to standardize a digital interface that would enable their digital musical instruments to communicate with one another (Durant 1990, 182). This protocol, labeled MIDI (Musical Instrument Digital Interface), was soon extended to several other digital instrument manufacturers as well. The computer was often used as a control center for managing these various digital instruments or MIDI processors.

The first computer-based software sequencer programs, which followed the introduction of home computers in the 1980s, could not operate audio sounds but could only record and control MIDI information—that is, the aspects that *affect* the sounds, such as which sound to use, its onset, offset, and duration, its velocity, and any pitch-bend or filter changes. However, up to the late 1980s, these programs were expensive and difficult to use, and the graphic user interface (GUI) was poor, often displaying little more than numbers in different columns. By 1990, the sequencer programs had become more user friendly, and within a few years the GUI dramatically improved as well, switching from numerical representations to visually *remediated* tools and equipment found in analog studios,[11] such as a pair of scissors to represent the ability to cut a track, and a virtual mixing console that resembled an actual analog mixing console, complete with input channels, faders, and panning knobs.

Already in 1989, the American company Digidesign (now called Avid Audio) had launched the first computer-based sequencer program, Sound Tools (the ancestor of Pro Tools), to offer audio recording (on two channels), thanks to dedicated processing hardware. Still, the final insertion of audio files into sequencer programs had to await better computer-processor power and increased random-access memory (RAM) capacity. The 1990s witnessed dramatic developments in computer technology as well as fully functional MIDI/audio-integrated DAWs; by the late 1990s, virtual instruments had even been integrated into the DAWs, in tandem with a wide variety of integrated software signal-processing effects that could be used in real time and whose quality was comparable to hardware equipment. As the capacities for computer storage, as well as processing and disk speeds, improved, the price of computers also dropped dramatically, as did the price of DAWs.

In fact, few of the components that make up the DAW are new. For example, the technology of digital recording can be traced back to the 1930s, when it was developed in the laboratories of telephone companies (Fine 2008, 1; Millard 2005, 346). The virtual mixing console, as well as the virtual instruments and signal-processing effects, are merely imitations or remediations of existing hardware equipment. The sequencer arrived already in the early 1970s with the EMS Synthi A, before the computer-based sequencer was introduced (Zeiner-Henriksen 2010b, 79). What *was* new about the DAW was that all of these technological inventions now existed in a single complete package that could be bought for a preposterously low price compared to its analog ancestors. During the 1990s, then, this fully functioning virtual studio made music production work a possibility for amateurs as well as professionals, thanks not only to its economic accessibility but also to its user-friendly interfaces. Andre Millard points to a *Rolling Stone* article from April 2003 that referred to the "Pro Tools Nation,"[12] "an indication of the unprecedented success of this software and the influential role it played in recording music" (Millard 2005, 386). Most contemporary popular music is produced with the help of a DAW program, often in combination with digital and analog hardware equipment.

A Digital Revolution?

The shift from analog to digital technology significantly influenced how music was produced. First and foremost, the digitization of sounds—that is, their conversion into numbers—enabled music makers to undo what was done. One could, in other words, twist and bend sounds toward something new without sacrificing the original version. This "undo" ability made mistakes considerably less momentous, stimulating the creative process and encouraging a generally more experimental mindset. In addition, digitally converted sounds could be manipulated simply by programming digital messages rather than using physical tools, simplifying the editing process significantly. For example, while editing once involved razor blades to physically cut and splice audiotapes, it now involved the cursor and mouse-click of the computer-based sequencer program, which was obviously less time consuming. Because the manipulation of digitally converted sounds meant the reprogramming of binary information, editing operations could be performed with millisecond precision. This microlevel access at once made it easier to conceal any traces of manipulations (such as merging tracks in silent spots) and introduced new possibilities for manipulating sounds in audible and experimental ways.

Manipulating sounds by programming digital information also facilitated the use of audio signal processing. For instance, compared to the analog tape delay, which involved the process of manually delaying a physical duplication of the audiotape from the original audiotape, or plate reverb, created by a vibrating steel plate, the digital equivalents involve simple mathematical adjustments to the sound information. By adding digital reverb to a sound, it could be sonically transferred from its original recording space to a completely different space. Digital reverb could create eerily realistic spatial simulations, but it could also add a presence to sounds that did not correspond to the spatiality of the exterior world.

Digital information could also be stored, which, among other things, made it possible to equip digital instruments, processing effects, and other digital equipment with presets—that is, already aligned settings of device parameters such as synthesizer patches, rhythm patterns on drum machines, and controller positions for various signal-processing devices. Producers could also store their various arrangements on their mixing consoles, which allowed for a trial-and-error approach and an accurate and instant re-creation of previously aligned settings. The latter ability also made it possible to execute abrupt transitions among several aspects of the mix simultaneously, in contrast to analog mixing, where changes had to be made manually, in real time.[13]

The fact that digital technology was able to produce perfect copies of sounds, without the slightest deterioration, also propelled the "loop aesthetic," especially in the genres of hip hop and electronic dance music, both of which were very important in the 1990s. While repetition had been a central structural feature of popular music for decades, it acquired new meaning in the digital age, when it came to encompass the repeated return of *exactly* the same chunk of music, over and over—in effect, musical clones. If the digital sampler of the 1980s pushed popular music farther into the frame of the "montage aesthetic," the computer-based sequencer programs of the 1990s and onward pushed everything a little farther still, in terms of freeing music from its spatiotemporal origins. One of the specific and enduring reasons for this is the visual environment that the sequencer programs present. The "editing board" interface or "arrange window" offered by the sequencer programs, in which editing operations—such as cutting, pasting, copying, merging, deleting, and moving sound sequences—are executed, displays audio and MIDI tracks that are arranged vertically in different channels and enfold horizontally across a grid-divided timeline.[14] This visual representation of the music now arguably influences how we compose it in the first place. For instance, using the cursor to drag and drop chunks of "music" across the timeline of the arrange window encourages us to think about music as consisting of bits and fragments that can be easily shuffled around, rather than as a continuous flow that evolves organically through time.

Digitally converted sounds can be treated very differently than analog sounds, and this, together with the unique sounds digital technology has introduced (which will be discussed later in this book), has affected how we make and think about music. It is, however, worth stressing that in suggesting that the digitization of technology has profoundly affected the aesthetics of popular music production, we are not thereby fomenting a "digital revolution" along the very lines cautioned against by Nick Prior:

"How does one avoid the overly uncritical and exuberant embracing of all things digital as revolutionary and transformative without suggesting that nothing has changed at all?" (Prior 2009, 82). In fact, digital technology has actually offered relatively few operations that are entirely *new*. And yet, because digitally converted sounds behave differently than analog sounds, this technology has clearly changed our approach to predigital technologies and practices.

Despite the obvious effects of digitization on popular music production, we do not promote a deterministic view of technology and in fact dispute Marshall McLuhan's famous claim that "the medium is the message," in the sense, at least, that technology, in a *determinate* way, constrains how we act or think in response to it (McLuhan 2010, first published in 1964). Still, though we do not regard the consumers of these technologies as passive or overly impressionable, McLuhan is right when he claims that technological artifacts can affect one's mindset, and that they tend to further certain operations at the expense of others. The notion that technology fundamentally influences society, which McLuhan claims, has been questioned by the social constructivist idea that technological artifacts are relatively neutral tools that are socially shaped. We are here confronted with the enduring debate about whether worldly objects have inherent properties (realism) or whether the "reality" of these objects is the result of social factors and processes (constructivism). Keith Grint and Steve Woolgar sum up the debate as follows: "Does technology … determine, or is it determined by, the social?" (Grint and Woolgar 1997, 21).

Here we take a position similar to that of Ian Hutchby (2001), who argues that technology does both. Instead of seeing specific technologies in terms of their "interpretive textual" properties or their "essential technical" properties, Hutchby sees them in terms of their *affordances*, drawing on the theories of James J. Gibson. One of Gibson's assumptions is that perception is always already intentional—those who "perceive and behave" are not processing masses of undifferentiated information but rather engaging with the environment to gather only that information that is meaningful given their purposes and context (Gibson 1986). In the context of the present discussion, then, a given technology affords particular possibilities to the consumer, in terms of enabling as well as *constraining* particular functions. Of course, as Gibson stresses when he introduces the term, "an affordance cannot be measured as we measure in physics" (ibid., 128). For one thing, it is relational: it may offer a function to one group of consumers but not to another. It might also offer one function in one context but not in another context. However, though an object's affordances might differ in these ways, they are not *freely* variable—there are things, in short, that an object does *not* afford, no matter what. In addition, the fact that an object's range of affordances is not fully and immediately available to perception does not mean that the object does not possess them; as Gibson explains, "The *affordance* of something is assumed *not* to change as the need of the observer changes" (Gibson 1982, 409; emphasis in the original). Thus, technologies are not empty or "open forms," as the radical constructivist position suggests, and they are not determinate, because a technological artifact might afford different things to different consumers. Hutchby's view of technological artifacts as in possession of different affordances for different consumers will inform our discussions of the effect of the digitization of technology on the aesthetics of popular music production, as well as our discussions of how technological mediation is experienced. While digital technology facilitates certain new operations, many music makers continue to use it as they did its predecessors. The position we take here thus complies with Hutchby's argument that "technological artefacts do not amount simply to what their users make of them; what is made of them is accomplished in the interface between human aims and the artefact's affordances" (Hutchby 2001, 453). With this book, then, we do not intend to imply some new form of "progress" in music making, which continues to thrive as much on tradition as it does on innovation. Instead, this book is written on the basis of the conviction that aesthetic and technological changes should always be understood in relation to one another. Musical and aesthetic dimensions are always pertinent to the process of understanding the cultural significance of technological change, and vice versa—aspects of technology have always played a role in changing the aesthetics of music.

Outline of the Book

The succeeding chapters delve into the musical material, using close analyses to discuss the novelty of the selected digital signatures, their use in the music production process, and their various contributions to the music's meaning. In our analyses we will address the following questions: (1) Why are the sounds and techniques in question often conceptualized as signatures of digital mediation? (2) How has their aesthetic potential been explored in the making of popular music? (3) In what ways have the digital production processes contributed to the experiential meaning of the music in our particular examples?

Chapter 2, "Making Sense of Digital Spatiality: Kate Bush's Eerie Collage," examines the signatures of digital reverb and delay in terms of how these processing effects have increased the possibilities for experimenting with spatiality in music productions. The point of departure is Kate Bush's "Get Out of My House" from the album *The Dreaming* (1982), with which we explore the way Bush and her coproducers deliberately used digital delay and reverb to suggest an aural spatiality premised not on the "real" world but on exclusively technological motivations that so evidently diverge from the sounds' "natural" acoustic behavior. Also of interest are the ways in which these opaque and otherworldly digital signal-processing effects might function metaphorically, in terms of either emphasizing the musical meaning or creating new meaning.

In chapter 3, "The Instrument Formerly Known as the Machine: Hyper-Accuracy and Sonic Richness in Prince's 'Kiss,'" we discuss the new sounds and procedures that were made possible by the combination of the MIDI protocol and digital synthesizers and samplers. These tools introduced a new palette of sampled and synthesized sounds and quantized timing to pop production in the 1980s, and they were often combined with the extreme clarity and precision in upper-frequency registers that were provided by digital reverb effects (discussed in chap. 2). In this chapter we discuss the sound of Prince's "Kiss" (*Parade*, 1986), pointing out the importance of new digital tools for the hyperreal character of this song.

Chapter 4, "The Rebirth of Silence in the Company of Noise: Portishead Going Retro," is primarily concerned with the characteristic *silence* of digital mediation. Here we also discuss how this silence has inspired music makers to revisit older technologies *because of* the unavoidable noises that accompany them, and how these noises (such as tape hiss or vinyl crackle) has enjoyed a rebirth of sorts in the age of their potential absence. As point of departure, we focus on Portishead's "Strangers" from the album *Dummy* (1994), which is characterized by a jarring contrast between the lo-fi sounds of predigital technologies and the silence and hi-fi signatures of digital mediation. Digital silence, which we are tempted to think of as "nothing"—as an absence of sound and, thus, an absence of meaning—is here vested with concrete musical import, in the sense that it is able to provide both aesthetic and emotional pleasure in the same way that the sounds of older technologies (now) do. We discuss how this juxtaposition of old and new represents a trend in which each is understood and enjoyed in light of the other, and how these sounds' meanings and functions are discursively dependent on each other.

In chapter 5, "Cut-Ups and Glitches: Los Sampler's and Squarepusher's Freeze and Flow," we seek to illustrate how the virtual cut-and-paste tool of the DAW offered an approach to music manipulation that was at once reminiscent of and different from the older practice of cutting and pasting using physical tools. Also of interest is how the sonic results of the cut-and-paste tool might be associated with, and used to match the sound of, malfunctioning digital technology—that is, glitches—and how these glitchy sounds might perceptually balance on the border between the music's interior and exterior. We discuss these digital signatures via "La Vida es Ilena de Cables" (Descargas, 2000) by Los Sampler's (Uwe Schmidt) and "My Red Hot Car" (2001) by Squarepusher, which both bear clear traces of cut-ups and glitches in their incomplete sounds, abrupt transitions between sound sequences, signal dropouts, stuttering rhythms, and other percussive effects.

Chapter 6, "Seasick Computers: Microrhythmic Manipulation in the Era of Endless Undo," sheds light on the particular rhythmic feels produced through manually or automatically manipulating the timing of rhythm tracks in digital audio workstations through "warping" of samples or by inserting temporal discrepancies between rhythmic layers. The resulting swaying or seasick feel was especially noticeable within African-American musical styles such as rap, neo-soul, and contemporary R&B early in the new millennium. We discuss this development through our analysis of selected songs from Snoop Dogg's path-breaking album *Rhythm & Gangsta* (2004).

Chapter 7, "Autotuned Voices: Alienation and 'Brokenhearted Androids'" discusses the effects of autotuning, which has been widely used in contemporary pop music. We show how this signature, which started out as a tool for pitch correction, soon developed into an artistic effect in its own right, so that its use was exposed instead of concealed. We start by discussing various hip-hop-related artists' use of autotuning as a means of portraying alienation and emotional distance, then present an analysis of the way in which indie rock artist Bon Iver uses an exaggerated and exposed autotuning effect in his a cappella song "Woods" (*Blood Bank*, 2009) that could be heard as a depiction of a form of hypernature. Finally, we discuss how an artist like Lady Gaga uses the characteristic morphed sound of human and machine to challenge traditional notions of femininity, as well as the trustworthiness of the human voice as an expression of authenticity and humanness.

Chapter 8, "Popular Music in the Digital Era," engages with the fundamental questions that are raised and explored elsewhere in the book. Pulling together our main arguments, we here speak to the significance of mediation in popular music, less in terms of its distribution and consumption than in terms of its very aesthetic.

Through combining a technical knowledge of digital music production with musical analyses, aesthetic interpretations, and theoretical discussions, we hope to provide insights into the ways in which the qualitative and quantitative changes brought about by digital mediation have profoundly affected popular music sound, while also exploring how the mediation in digitally produced popular music is *experienced*—how it affords a wide range of meanings. In this way, we hope to contribute to the ongoing scholarly discussion regarding the role of technological mediation in the aesthetics of popular music, as well as the general understanding of music technology in use.

2 Making Sense of Digital Spatiality: Kate Bush's Eerie Collage

I don't miss out on normality ... I'd rather hang on to madness than normality anyway.
—Kate Bush[1]

One of the significant changes in popular music production brought about by digital technology is a new means of fabricating musical spatiality. Musical spatiality refers both to the sonic locations of the sounds within space and to the sonic design of the space itself. In this chapter, we are most interested in the latter. While music makers have been fascinated by and have experimented with this form of musical spatiality for years, the ease of molding virtual spaces in music with digital reverb and delay effects has given rise to new approaches that reinvent spatiality and sometimes accommodate several distinct reverb and delay effects within the same track.

While digital recording did not become standard in recording studios until the early 1990s, digital delay was introduced to the commercial market already in the mid-1970s, and digital reverb was introduced in the late 1970s and early 1980s. "Delay" is here synonymous with "echo" and refers to discrete and discernible reflections of a sound, whereas "reverb" (or "reverberation") refers to reflection patterns that are so numerous and dense that they cannot be distinguished perceptually. The sound processing effects of both delay and reverb imitate the patterns that are produced when a sound is reflected by surrounding walls or other obstacles in actual spatial environments, but they also facilitate spatial designs that differ from natural acoustic designs altogether, such as gated or reverse reverb, which will be discussed in detail later in this chapter.

We will begin the present discussion by pointing out a few crucial ways in which our experiences with natural acoustics inform our understanding of musical spatiality, drawing upon key aspects of Denis Smalley's theory of source bonding. To situate digital reverb and delay effects in their historical context, we will then describe how the means of fabricating musical spatiality developed throughout the twentieth century, paying attention to what is new about the digital tools in particular. We will also observe that, from very early on, we can recognize two different paradigms of musical spatiality, one that aims at simulating the ways in which sounds behave in actual spaces, and one that cultivates spatialities that do not exist outside of technologically mediated environments. We are particularly interested in the latter paradigm, because it tends to expose its mediation in an opaque fashion.

As a case study informing our analysis of digital delay and digital reverb as artistic tools and as qualities intended to impact listeners, we will examine Kate Bush's "Get Out of My House" from her 1982 album, *The Dreaming* (EMI). Produced relatively soon after digital reverb and delay were introduced, the music clearly indicates the fascination of Kate Bush and her coproducers with these new effects. In our analysis, we will discuss how the track's sonic design allows the music to be experienced as surreal, because its musical spatiality clearly differs from any actual physical environment. In our conclusion, we will further observe that this sense of the surreal generally becomes relatively naturalized as we become more familiar with the sonic design that evokes it. We will also discuss how musical spatiality might be used to either supply new meaning or emphasize a meaning that is already conveyed by other musical or lyrical aspects of the track, such as generating or underlining various emotions, atmospheres, and personas.

Conceptualizing and Fabricating Spatiality in Music

In his investigations into the listener's perception of the spatial image of electroacoustic music, Smalley argues that sounds are generally "source bonded" in the sense that, as human beings, we have a *"natural* tendency to relate sounds to supposed sources and causes, and to relate sounds to each other because they appear to have shared or associated origins" (Smalley 1997, 110; emphasis in the original). A listener might therefore be expected to hear recorded sounds as signs of actual spatial environments, because people in general are used to hearing sound as signifying space.

Accordingly, music makers also often think in terms of creating sonically "virtual spaces" within the music.²

In an enclosed physical space, such as a room, sound travels in all dimensions and bounces around as it meets the surfaces of walls, the floor, or the ceiling in turn. The multiple (and multiplying) sonic reflections of the sound gradually weaken as the air and the surfaces absorb them, until they die out entirely. If a sound hits its first surface (preferably a hard one) after fifty to eighty milliseconds (depending on the sound itself), its reflections will be audible as distinct, separated sounds—what we refer to as echo or delay (Rossing, Moore, and Wheeler 2002, 528). Enclosed spaces that produce echo might include large empty buildings or wells; outside spaces might include neighboring mountains or tall concrete walls. More common than the echo effect (in nonmusical circumstances) is, of course, the effect of reverb—that is, when a sound hits a surface immediately and promptly hits many others as well (a ping-pong effect), producing multiple echoes that are so dense that the reflections cannot be distinguished from one other.³ A sound deprived of any acoustic reflection sounds unnatural. The size of the room determines the temporal duration between the source and its initial decay (often referred to as the "predecay time"), as well as the duration of time before the sound's subsequent rapid and complex reflecting pattern dies out (the "reverb decay time"). The texture of the surfaces in question (concrete, glass, or wood, for example) determines the *reflectivity* of the sounds—that is, the extent to which they are absorbed—which affects the loudness and frequency response of the reflections, as well as the reverb's decay time (Pohlmann 2000, 633).⁴ When we hear recorded music, we recall the sounds of different complex acoustic reflection patterns, and in this way sounds can function as signs or bearers of actual spatial environments.

In addition to the sonic design of the space itself, which is primarily constituted by reverb and delay, musical spatiality, as mentioned above, also refers to the sonic locations of the sounds within space, or the spatial organization of the sounds within the music production—that is, within the space between the speakers. This aspect of sonic spatiality is what inspired Allan F. Moore to develop his "sound box" model.⁵ He suggests that a sound's frequency register might be conceptualized as its placement in the vertical dimension of the sound box; its location within the stereo image might be conceptualized as its placement within the sound box's horizontal dimension; and its various signal-processing effects, such as delay and reverb, might contribute to its conceptualization as placed in the foreground, middleground, or background of the sound box. What is clear is that the sound box is not a description of the virtual sonic space per se but a music-analytical tool that can be used as a matrix to map the spatial placement of the different elements of a mix (Moore 2001, 121).[6] Several sound engineers have also described approaches to mixing that recall Moore's sound box (see, e.g., David Gibson's *The Art of Mixing* [2005], and Georg Massenburg's method as discussed by Albin Zak in Zak 2001). The perceived placement of sounds within the music (and the sound box) is determined by production parameters including relative loudness, microphone placement, dynamic compression, frequency content, ambience, and stereo volume. In our analysis of "Get Out of My House," we will take several of these aspects of musical spatiality into account.

The technological means of fabricating the virtual space conveyed by a musical recording, and of reproducing the sonic atmosphere of actual spaces, have developed since the arrival of the phonograph itself. In the early days of recording, the virtual space was determined solely by the ambience of—and the instruments' placement within—the specific acoustic space holding the performance; few tools existed for engineering anything more. However, owing to the phonograph's inability to capture soft sounds and to re-create all frequencies of sounds, the acoustic atmosphere on a recording was very different from that of the recording space. Electromechanical recording, which became the standard recording technology starting around 1925, enhanced the possibilities for amplification significantly, thanks to its conversion of sound into electric currents. Engineers could now capture much softer sounds, allowing musicians to move further away from the microphone and thus strategically occupy more space in the recording environment; in addition, the microphone absorbed more of the room's general ambience as well. The next step toward taking more control of the virtual space of the recording was to carefully choose the architectural frame in which one recorded. In the 1940s, engineers even built "echo chambers" to create special reverb effects. The engineer would then place the sound source (an actual performer, or a loudspeaker playing a recorded sound) within this chamber, together with a microphone to pick up the sound and its reflections (Doyle 2005, 27).[7]

Starting in the early 1960s, when the magnetic tape recorder had become the standard recording medium and stereophonic sound and multitrack recording were well underway, individual musical parts could be recorded separately, without bouncing them onto a single track, and then processed and placed within the mix independent of any other sounds. Engineers created the spatial environments of sounds either by recording the sounds' natural reverb in different spaces (rooms or whole buildings) or by using the new technology of "plate reverb." The principle behind the latter was that the reverb resulted when the amplified electrical currents of the sound signal generated vibration in a suspended thin steel plate. Even cheaper technologies based on the same principle used a spring instead of a plate—thus, "spring reverb." When the artificial reverb effect was to be applied, the sounds would be recorded in a dampened room so as to minimize natural reverb and derive a relatively dry sound, to which the plate or spring reverb could be added.[8] Musicians and engineers at this time also started experimenting with the tape path to create an artificial echo or delay: by adding an extra playback head to the recording machine and combining a reel-to-reel tape with a looped tape sequence, the signal on the reel-to-reel tape could be recorded on the tape loop while it played back. When the tape loop ran through the playback head itself a few seconds later, the sound that had just been heard was repeated. The tape loop then entered the erase head to begin the process again (the length of the tape path from the recording head to the playback head determines the delay time).[9]

Since the digitization of music-related technology, it has become possible to create artificial reverb and delay in more complex but also more controllable ways. Digital delay is produced by storing converted sounds in short-term memory before output following a delay dictated by the user. Digital reverb has a much more complex design; it is created via various algorithms or mathematical formulas that alter the numerical values of the digitized sound signal to simulate all of the different parameters at work in natural reverb.[10] It is built up in a similar way to natural reverb, and thus consists of a predecay time, which is the time preceding the sound's initial decay; early reflections, which are relatively discrete; and later, denser reflections. The early and later reflections are often referred to as reverb decay time, or "reverb tail"; see figure 2.1.

Digital reverb not only can reconstruct the timing of these different "natural" reflections but also can capture the changes in volume and frequency.

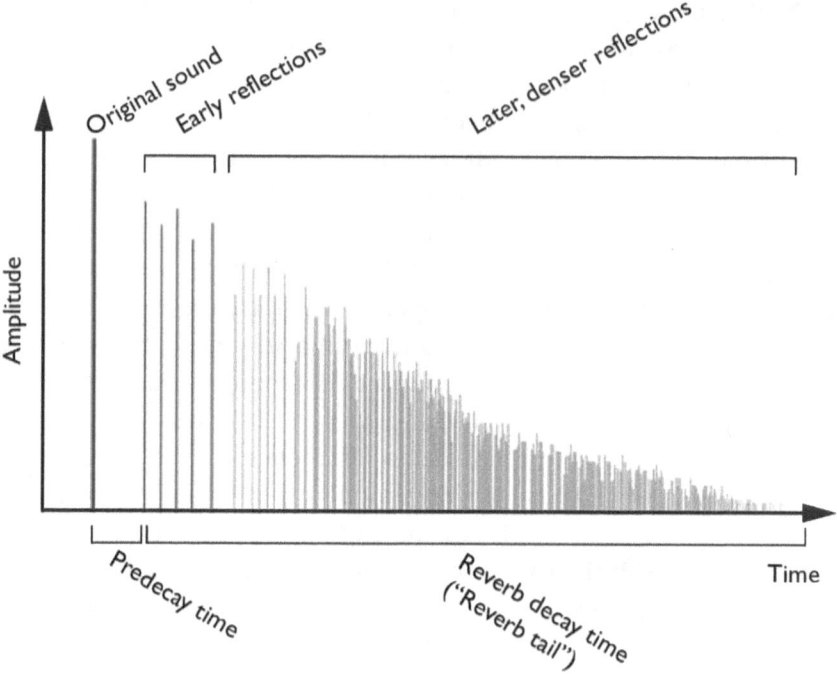

Figure 2.1
The figure depicts a typical reflection pattern of reverb, and the terms used to describe its different aspects.

With digital reverb's superior access to every parameter of the natural effect, producers had more control than ever before. Since the volume, frequencies, placement, and duration of the reproduced sound reflections could be altered, it was now possible to produce musical spatialities that were more realistic than before (for instance, plate reverb does not have a predecay time and thus sounds less natural than digital reverb), as well as producing sonic spaces that were larger or smaller than life, or different from life in other ways, as we will discuss later in this chapter. Moreover, digital reverb was completely clean, in contrast to plate reverb, for example, which added a characteristic metallic ring to its sounds.

From Doyle's analyses in *Echo and Reverb: Fabricating Space in Popular Music Recording, 1900–1960* (2005), it is clear that, very early on, there were two prominent production alternatives for creating musical space. One was to produce a virtual spatial environment that sonically re-created the space of the original recording session, or any other "worldly" space. The other was to produce a virtual spatial environment with features that clearly differentiate it from any familiar actual space. Doyle points to certain music recordings as far back as the late 1940s and early 1950s in which the virtual spaces reveal a "strong sense of 'manufacturedness,'" as he puts it (see Doyle 2005, 143–162). In Patti Page's "Tennessee Waltz" (1950), for example, the electric guitar goes from reverberant to dry in only two bars in the introduction, while Speedy West and Jimmy Bryant's "West of Samoa" (1954) alternates between "dry" and "wet" verses, which, according to Doyle, "serve[s] to cast the listener in and out of a mysteriously exotic, more than a little threatening soundscape" (ibid., 156). These pioneering early attempts at spatially "surreal" sound in predigital music set the stage for the manifestly greater possibilities inherent to reverb in the digital era.[11]

In the present study, we are particularly interested in the use of reverb and delay effects that diverge from natural acoustic reflection patterns, because they are most likely to be noticed by the listener. Those effects that mimic the real world, in other words, are most often experienced as transparent, whereas those that diverge make obvious the fact that their sounds have been split off from their original spatial setting and remade as something else.

During the late 1970s and early 1980s, when digital delay and reverb first came onto the scene, musicians and producers often exaggerated the new functions they offered, lending certain recordings of the time a rather pronounced sonic trademark. Kate Bush's "Get Out of My House" is among those recordings, and we will analyze it here in terms of its exaggerated suggestion of spatial environments that differ from actual environments. Yet we will also demonstrate Doyle's further point that, in terms of music like this, "while flirting with the supernatural," such "surreal" virtual spaces seek nevertheless to "trigger in the listener mental images of coherent, 'imaginable' physical spaces" as well (Doyle 2005, 8). That is, musical spatiality has a tendency to point the listener toward a real-world physical phenomenon even as it acts to undermine that reality.

A House of Surrealistic Spaces

Born in South East London, Kate Bush was in her late teens when she was reputedly discovered by Pink Floyd guitarist Dave Gilmour, who found her a contract with EMI that led to her debut album, *The Kick Inside*, in

1977. She has since managed a long and successful career in popular music thanks to (or in spite of?) her unique and unconventional musical and lyrical ideas, her idiosyncratic singing style, and her involvement in the production process of her music. Her vocal performances are characterized by their extremities, physically and emotionally, and her music is eclectic and experimental, blending traditional rock instruments with ethnic instruments. Most of all, she obviously embraces technological innovations, including electronic instrumentation, processing effects, and experimental production techniques.

"Get Out of My House" is the last track on her fourth full-length album, *The Dreaming*, which caused much controversy for its aggressive realization of Bush's last lyrical line in the album's fifth track, "Leave It Open": "We let the weirdness in." One reviewer, from the music blog *Glorious Noise*, describes the album as showing "a young woman, manic with ideas and creativity, throwing caution to the wind and delivering an off-her-rocker masterpiece that very few artists have ever had the courage to make before or since" (Totale 2009). However, according to Kate Bush herself, the process leading up to *The Dreaming* involved a very conscious alteration in her production techniques. Among other things, she exchanged the piano for the drums as her main tool in the songwriting process: "I felt as if my writing needed some kind of shock, and I think I've found one for myself. ... The piano, which is what I was used to writing with, is so far removed from the drums, so I tried writing with the rhythm rather than the tune" (quoted in Shearlaw 1981, 6). Another important aspect of the album's renowned "weirdness," of course, is the extensive technological mediation involved in its production.

According to an interview that *Keyboard* journalist John Diliberto did with Kate Bush in 1985, *The Dreaming* was the first album that she coproduced. She took the producer's seat, she said, because she wanted to integrate technological mediation into the musical composition itself: "By the time the second album was finished, I knew that I had to be involved [in the production process of the music]. Even though they were my songs and I was singing them, the finished product was not what I wanted. That wasn't the producer's fault. He was doing a good job from his point of view—making it sound good and together. But for me, it was not my album, really. ... The more I get involved in the production, then the more I'm going to get exactly what I can out of it. Therefore, it automatically becomes a more

Making Sense of Digital Spatiality

demanding and personal project" (quoted in Diliberto 1985). With a range of digital equipment at her disposal and the guidance of established producers including Hugh Padgham, Nick Launay, Haydn Bendall, and Paul Hardiman, Bush used *The Dreaming* to experiment with production techniques and exploit obvious signatures of digital mediation.

In the same interview, Bush also indicates that they used the digital Quantec QRS "room simulator," which was introduced in 1982—the year that the album was made and released—to add reverb to the different instrumental and vocal sounds on *The Dreaming*.[12] She was attracted to its improvements around producing spatiality in music: "We have a room simulator called a Quantec, which is my favorite. It would be lovely to be able to draw the sort of room you wanted your voice to be in. I think that's the next step" (quoted in Diliberto 1985). This would prove true, as several ensuing software reverb effects would offer a graphic interface actually depicting the rooms that the reverb settings simulate.[13]

In "Get Out of My House," the digital reverb and delay present an otherworldly musical spatiality. One reason for this is the distinctive nature of the reflection patterns that Bush applies. The other reason is the track's combination of several different virtual spaces at the same time. While each of these juxtaposed spaces could be heard to simulate an actual space, the sonic collage they comprise could never be experienced in real life.[14] We will first look at how Bush used these processing effects on the instruments and vocals of the track, then at how listeners might experience the spatiality they set up as diverging from actual spatial environments. Finally, we will discuss how the spatiality might be underpinning the meaning of the song. In addition to describing the reflection patterns of the sounds and the ways in which they diverge or comply with the reflection patterns of actual spaces, we will also place the sounds and their spatial environments within the analytical "sound box"—that is, the *abstract* space between the speakers. In our attempt to place the sounds specifically on the depth axis of the mix, we will draw upon Edward Hall's classification of the four "distances" he finds most relevant to social situations in his study *The Hidden Dimension: Man's Use of Space in Public and Private* (1969). The first, "intimate distance," extends up to eighteen inches from a person; the second, "personal distance," extends from eighteen inches to about four feet. The third, "social distance," extends from four to about twelve feet, while "public distance" extends from twelve to twenty-five feet or more (Hall 1969,

107–120). While Hall applied his classifications to human territorial behavior, we will apply them to the apparent physical distances implied within music from the listener to the sounds.[15]

"Get Out of My House" starts with a guitar chord, drenched in reverb, that fades in and then descends in pitch, heralding in turn the instruments that will form the basic accompaniment of the track: the electric guitar, the piano, and the drums (the electric bass guitar does not appear until the first verse). Following the sustained guitar chord, we hear a melodious, heavily compressed electric guitar riff that consists of three played tones per bar but sounds like six tones. This is because a digital delay repeats the played guitar tones after a pause of an eighth note, one time each, as we can see in figure 2.2.

As mentioned, analog tape could also produce artificial delay effects, but not as well or as flexibly. The quality of the digitally delayed sound, for one thing, does not degrade at all, whereas a delayed analog signal always suffers from generation loss (Watkinson 1999, 115). In addition, the digital delay time can be specified within milliseconds but also synchronized to note measures (such as eighth notes), which made it much easier to match the delay effect to the tempo of the music. (The delay time for analog tape is determined solely by the length of the tape path from the recording head to the playback head, which could only be controlled by choosing among multiple mounted playback heads or by altering the position of a playback head mounted on a slide.)[16] Modern digital delay can mimic all of these shortcomings, of course, but more important to Kate Bush and her ilk was its ability to transcend them.

In "Get Out of My House" Bush demonstrates this interest in the new and unique delay effect of sonic clones with precise timing. The delay suggested by the guitar sound does not simulate an actual spatial environment;

Figure 2.2
Transcription of the guitar riff of the introduction (a) with delay (the delayed sounds have stems pointing down), and (b) without delay.

a natural echo would reveal lower amplitude than the original signal, and less presence in the upper frequency range. The volume and frequency ranges of the delayed sounds in "Get Out of My House" are identical to the original guitar sounds, drawing renewed attention to the fact that the riff consists of three played sounds that are delayed rather than six played sounds, which could never be matched so precisely. The design of this guitar riff, then, clearly belongs to the digital era.

This is not the only digital signature, however: the assortment of sonic environments itself gives the game away. The piano, which is played in a minimalistic, chord-based staccato style, sounds as if it is situated within a small, dampened room, which might be the result of where it was actually recorded or of a digital reverb effect designed to simulate the same. The electric bass guitar, which propels the track's groove with a riff based on eighth and sixteenth notes, sounds dry, as if the bass's output cable had been plugged directly into the mixing console, thus avoiding the natural reflections of the recording room altogether. Contrary to both the piano and the electric bass guitar, the minimalistic but forceful drum part, which sounds like slamming doors (presumably in tandem with the lyrics about slamming doors), seems to have been played in a large and empty room. There is, however, an odd twist to the sound even here: while the reflection pattern of an actual large and empty room involves a long reverb decay time (the period as the reflections fade out), the reverb on the drums in "Get Out of My House" is cut off after only a few milliseconds. Instead of the sound fading into a big space, the space disappears altogether, rendering the "big" sound suddenly dry. The effect is almost surreal, as Zak points out in his description of gated reverb (see more below) as well: "We are immediately taken from the acoustic world as we know it into a strange soundscape of unknown dimensions where sounds behave in unfamiliar ways and the air itself is controlled by machines" (Zak 2001, 80). The latter feeling emerges from the incongruity between the reverb pattern of this virtual sound room and our familiarity with reverb patterns in actual enclosed environments: reverb usually persists after a sound has stopped, whereas here, the sound persists after the reverb has stopped.

This particular drum sound is referred to as "gated reverb" because it was first achieved with the help of a "noise gate," a processing effect that reduces or eliminates any sound signal that appears beneath a threshold limit set by the producer. To achieve gated reverb using analog technology, one microphone was placed close to the sound source while a second microphone was placed further away (to capture room ambience) within a recording environment chosen for its huge amount of reverb. During playback, the highly compressed ambience track is sent through a noise gate set to a high threshold level, which removes the reverb tail of the sounds, while the relatively dry close-microphone signal plays normally. Hugh Padgham, one of the most sought-after British producers of the 1980s, is famous for pioneering this effect using analog technology, first on the drums on Peter Gabriel's third solo album from 1980 (*Peter Gabriel 3 [Melt]*, Charisma/Mercury/Geffen) but more famously on the drums on Phil Collins's 1981 hit "In the Air Tonight" (*Face Value*, Virgin/Atlantic).[17] Padgham also contributed to the production of Kate Bush's "Get Out of My House," which might explain the choice of drum sound here.

Though the gated reverb could be produced with analog technology, it is more strongly associated with digital technology, because it is so much easier to create there. Whereas an analog gated reverb effect requires an actual large recording room (for "In the Air Tonight," engineers not only built such a room but also covered all of its surfaces with stones [Cunningham 1998, 325]), the digital effect involves only algorithms. Instead of juggling microphone placement and manipulating playback equipment, in addition, engineers could simply reprogram the digital information to remove the reverb tail.[18] Digital reverb effects soon arrived with a "gated reverb" preset, and it was used on countless recordings during the 1980s, particularly with the drums, as Mark Cunningham points out: "Ambient, gated drum sounds proudly reigned throughout the Eighties to the point when artists and their producers stressed the importance of such qualities to almost obsessive levels" (Cunningham 1998, 322). Although it is not *unique* to the digital era, then, the sheer frequency of it (a *quantitative* change) makes this reverb effect a signature of the digital.

A third typically digital effect that reveals the spatiality's manufacturedness in "Get Out of My House" can be heard on the vocal passages between 00:58–01:00 and 2:14–2:16 (min:sec)—namely, reverse reverb, that is, a reverb pattern performed backward (as the source sound is played forward). An engineer could also achieve this effect with analog technology by placing a tape bit with recorded sounds on the tape recorder backward, then adding reverb to it as it plays in reverse. When the engineer then flipped the tape on the recording machine so that the sound played forward again, the reverb would be playing backward. However, this effect was also much easier to achieve by reprogramming digital information, and it became much more common in that context.[19]

In addition to its use of reverb patterns in ways that diverge from any naturally produced acoustic reflections, "Get Out of My House" represents a spatial collage. William Moylan points out that our spatial "imagination" when listening to music is influenced both by the sonic characteristics of each sound and by the overall sound created from those individual sounds. Writing from the perspective of studio production, Moylan also points out that producers are usually conscious of both of these levels of sonic spatiality: "In current music productions, it is common for each instrument (sound source) to be placed in its own host environment. This host environment of the individual sound source (a perceived physical space) is further imagined to exist within the perceived performance environment of the recording (space). This creates an illusion of a *space* existing *within* another *space*" (Moylan 2002, 176–177, emphasis in the original).[20] Each individual sound occupies a subspace within the music's all-encompassing spatial environment. Smalley similarly observes that the "holistic" space of the music comprises "zoned spaces," and possibly also subzones (Smalley 2007, 37). Even when a musical production is in fact a montage of zoned spaces, it is not always heard (or intended to be heard) as such. But sometimes it is precisely the point to generate the effect of superimposed spaces. Smalley describes this as a *spatial simultaneity*—that is, an occasion when "you are aware of simultaneous spaces" in the music (Smalley 1997, 124). When we listen to music that suggests a spatial collage, we do not draw upon any given experience with a particular space but are rather forced to attempt an awkward synthesis of a number of such spaces. As we project these previous "real-world" experiences onto a single virtual environment, we hear the music in question as unnatural or surreal and opaque, because it clearly signifies a spatial environment that could never occur in a real, physical, technologically unmediated environment.

While this effect is not unique to digital reverb as such—it could be recreated by capturing different analog sounds in different spaces and emphasizing this diversity in the mix[21]—the ease of creating it digitally has turned popular music in the general direction of the sonic collage to an unprecedented extent. Key to this development were the digital reverb's presets (preprogrammed digital algorithms), which simulated different spatial environments and were labeled after the room or the reverb they tried to simulate (such as "cathedral," "large hall," "medium hall," "large room," "small room," or "plate reverb"). These presets allowed one to experiment with a wide range of reverb at the push of a button (and one could further adjust the selected reverb as desired). By facilitating the creation of spatiality in the music, digital reverb encouraged the collage aesthetic that characterizes "Get Out of My House," especially its vocal sounds. In what follows, we will describe the different spatial environments of that vocal collage and explore how these various spaces are used to underline what is being communicated through the vocal performance.

A House of Eerie Spaces

As Peter Doyle (2005) and Serge Lacasse (2000) point out, musical spatiality evokes not only actual spaces but also *metaphorical* ones. For example, a certain kind of reverb might indicate both the emptiness of an actual physical environment *and* the "emptiness" of despair (see Lacasse 2000, 179). In "Get Out of My House," the sonic collage brought about by the production's copious digital delay and reverb effects emphatically influences the listener's interpretation of its musical meaning.

In an interview, Kate Bush revealed that "Get Out of My House" was inspired by Stephen King's horror novel *The Shining* from 1977, which was adopted into a film of the same title, produced and directed by Stanley Kubrick, in 1980, two years before Kate Bush released *The Dreaming*. The novel is about a man named Jack, who, in an attempt to escape his troubled past and start over, takes a job as caretaker at an isolated resort hotel and moves there for a snowbound winter with his wife and son. Trapped there, in effect, Jack is then possessed by a supernatural force or ghost that lives in the hotel, and his ensuing detachment from reality culminates in his attempt to kill his wife and son. A reprint (available online) of Kate Bush's self-authored article in issue 12 of the *Garden* in 1982 indicates that she related the story of *The Shining* to the atmosphere she wanted to capture in "Get Out of My House": "*The Shining* is the only book I've read that has frightened me. While reading it I swamped around in its snowy imagery and avoided visiting certain floors of the big, cold hotel, empty for the winter" (Bush 1982).

The many rooms, corridors, and doors that appear both in the sound production/effects and in the lyrics of "Get Out of My House" could

Making Sense of Digital Spatiality

certainly evoke associations with the frightening, desolate spaces of the hotel in *The Shining*. The track's powerful and complex musical spatiality could also work metaphorically, however, expressing other *Shining*-related sentiments around a person haunted by her past, or a person chased after or invaded by another person/being, or a person struggling with an inner conflict. Through her theatrical vocal performance, Kate Bush captures several characters and a huge range of emotions during the track, each of which is underlined by a digitally rendered virtual spatiality.

The first character Kate Bush captures, who appears between 0:13 and 0:46, conveys fear and misgiving as she sensitively delivers the lyrics suggesting that "she," or the main character of "Get Out of My House," is frightened by a man. The impression that the follower is a male follows from the masculine-voiced interjections as well as a dialogue between Bush and a man toward the end of the track; of course, this man might symbolize any person, demon, or otherworldly phenomenon that she fears. The lyrics associate the fear with the other's potential for instability, either mental or physical: "When you left, the door was slamming / You paused in the doorway / As though a thought stole you away / I watched the world pull you away." She is frightened so she "run[s] into the hall" and "into the corridor." She is looking for a "door in the house" as she listens to "the lift descending" and sees "the hackles on the cat standing." She finds the door and "lock[s] it" with her key.

Applied to this voice is a "slapback" delay with only one repetition, which is not very distinct but still contributes to the overall sound of the voice. Slapback is basically an echo with a short yet perceptible decay time. Whereas, for instance, the slapback echoes of the lead vocal in Elvis's "Mystery Train" (1953) were quite distinct, those applied to Bush's lead vocal here appear in a gray area between what can be discerned as actual slapback and what might instead be heard as fattening the sound. In addition to this particular slight delay, this vocal has a reverb whose reflection pattern suggests a relatively large, empty hall full of hard surfaces, such as concrete or stone. The reverb is in stereo, which gives the impression that this large concrete hall in fact surrounds us. The lead vocal reinforces this impression. It sounds as if it were recorded or "dubbed" three times, whereby one of the takes is placed in the middle of the horizontal axis of the sound box, or mix, while the other two are placed left and right (approximately forty-five degrees either way). The vocal sounds as though it is within our "personal" distance—that is, according to Hall's categorization, between eighteen inches and four feet away—making Bush seem very close to us. These reverb and slapback effects in turn play neatly into the scary ambience of Bush's lyrics and vocal performance. The slapback echo manages to make the otherwise resolute vocal sound broken or vulnerable, as if the character were shivering. Because the vocal is dubbed and spread across the stereo field, the fear it expresses surrounds, even immerses, us. Moreover, the "hard" reflection surfaces of the voice's reverb seem to evoke the deserted hotel described in the novel *The Shining*, for example, or otherwise give the impression that this person is located in the sort of large, empty building or storehouse that is so often exploited in horror movies and crime TV series.

The drumbeat of this section, which is processed with gated reverb and sounds like doors slamming, also reinforces the impression of a large, empty space. More precisely, it suggests the resonance of a large hall suddenly reduced to nothing, as if a door has been closed (though the sonic result here is in fact very different from what it would be in reality). Interestingly, one of the "doors" within each bar—the eighth note before the downbeat (or the "one")—does *not* reveal gated reverb but instead a relatively dry sound, and it is panned hard left. The result of this discrepancy is that while the other doors seem spectral as they slam in the "background," this door sounds like it is very near to us, and thus very concrete or even real. It is a jarring, unsettling alternative to an already unpleasant and eerie atmosphere. The persistent thumping and minimal pattern of the drums also suggests the stopping and starting of a beating heart, while the electric bass guitar's monotone riff on eighth and sixteenth notes suggests a racing pulse that we feel as much as hear, thanks to the dry sound of the strings.

At 00:47, we are introduced to Bush's second vocal character; this persona's voice is drawling and nasal, the latter association strengthened by a volume boost on the frequencies that produce nasality (1,000–2,500 kHz). Situated within our "social distance" (four to twelve feet away), she speaks to us with an exaggerated cockney accent: "I am the concierge chez-moi, honey / Won't letcha in for love, nor money." Up to now, Bush has been telling us a story about someone she fears (using an accusatory second-person "you"); here, though, the change of character, from frightened to more defiant and even nonchalant, seems to suggest that we now see her from outside, the way the man (or the "you") sees her. Again, this change of "scene" is emphasized by the use of a different musical spatiality. The sound's short predecay time (the time before the sound reaches its first obstacle) and short reverb decay time suggest a small room with surfaces that absorb most of the reflections that constitute the reverb.

A male vocal interjects upon the cockney vocal part, demanding "Let me in!" (0:54–0:55). If she had thought that she was protected from the man by a locked door, as the lyrics imply, his voice is now right beside her. It is placed hard left in the stereo field, and its predecay time and reverb decay time are so short that the listener might not even notice the reverb—the male voice sounds almost completely dry and appears to be situated within our personal distance. This house, then, begins to appear metaphorical rather than actual—perhaps it represents her body or psyche, and the keys and locked door are symbols of her inner struggle against the intruder. She ignores him, regardless of the apparent proximity: "I'm barred and bolted and I—" At this point in the sentence, which concludes "—won't let you in," the voice is drenched in the reverse reverb described earlier, as if she is drowning, either in her own fear or even in real life, if the man has in fact attacked her. This voice is helpless and diffuse, yet we sense rage as well, which is boosted by a distortion effect that is added to the vocal, as well as dubbed backing vocals shouting furiously from elsewhere in the track's eerie spatiality: "Get out of my house!"

Between 1:03 and 1:30, Bush takes on a third character who sounds at once more resigned and dejected but also somehow resolute as she sings: "No stranger's feet / Will enter me / I wash the panes / I clean the stains away." The lyrics suggest that this new voice turns the narrator toward her own inner space, in contrast to the outward focus of the first and second characters and their spatial settings, and this impression is further strengthened by the reverb. With no slapback, no nasality, and no hard reverb reflections, this vocal sounds much cleaner and more like an "everyday voice" than her previous incarnations. This vocal part has a relatively long predecay and reverb decay time, suggesting that the sound has traveled a significant distance before being reflected by the room's surfaces—hence, from a room of considerable size. However, since the mix is so dense with other sounds, these reflections virtually disappear in the periphery of the mix. While reverb often cause sounds to seem to be further away, this vocal's relatively long predecay time belies that sense; because the time before the sound meets its first obstacle (the initial decay) is rather long, the sound is not defused by its own reverb but remains clean and clear so as to sound very present and "up front." Similar to the slapback vocal, then, this vocal is placed within our personal distance, but not in stereo—instead, it appears to be placed directly in front of us. The character has left the house built sonically of stone and entered her/our inner, more intimate and calmer space.

While these different spaces are not necessarily eerie in and of themselves, they underline the wide emotional spectrum that the main character is experiencing throughout her inner struggle with the intruder. They do so both by affecting the sound of the voice in different ways and by suggesting various physical spaces, as well as different narrative perspectives. As the music progresses, the sections with the frightened "slapback character," then the concierge character, then the resigned but resolute character cycle through in the same order but build emotionally, and the atmosphere becomes even more eerie and unpleasant.

The track ends with a melodious dialogue between the man and the woman (3:13–3:48). The man starts by singing softly and almost gently: "Woman let me in / Let me bring in the memories / Woman let me in / Let me bring in the devil dreams." He sounds less demanding and threatening than before, partly thanks to the light guitar melody behind him. Yet there is unease, obviously, in his promotion of "devil dreams." The woman answers him politely but firmly: "I will not let you in / Don't you bring back the reveries / I turn into a bird / Carry further than the word is heard." He persists, in a voice saturated with reverb but placed within our intimate distance range. In this case, a relatively long reverb decay time preserves the sound's definition, but it is also clear that the vocalist stood quite close to the microphone during recording, and the voice has been further processed with a compressor. Every breath, swallow, and random throat noise is audible. Finally, the male voice is dubbed (recorded two times), and a take is placed hard left and hard right, respectively, so that he sings directly into our ears. While we know that he is addressing the woman in the music's narrative, we are made very aware that we have *become* her as we are sonically overwhelmed and invaded.

The woman's voice is just as present as the man's, however, so we become him as well, in a sense, as we listen to her. Though she sings calmly, her voice sounds fragile and brittle: "I'll turn into a mule," she warns. His voice becomes sinister as he responds, "Let me in!" This opening of hostilities is reflected in his voice, now panned hard left and almost free of acoustic reflections, as if he has departed the music itself and now stands right next to us.

While at first the "mule" seems to suggest only her resistance, soon it becomes something more, as she starts literally braying or bellowing in agony between loud intakes of breath. This mule sounds abused or at the very least overtaxed. After a time, the man takes over the braying and Bush's voice disappears for the rest of the track. The man's performance of the "mule language" has a very different character from Bush's: there is triumph rather than agony or fear, and the heavily compressed, intermittent breathing sounds that are added atop the braying (and within our intimate distance) sound almost erotic. This conclusion, then, brings with it very troubling undertones of sexual assault or a very painful spiritual possession. Alternatively, however, we might hear the "male" roaring as the woman's own battle cry, as she claims yet another (victorious) voice after having scared her intruder away. The track ends with a male choir reciting syllables from the Indian drum language Kannakol: "Dha dhin; dha ga ta; ta ka dha dhin; ga ta; dha dhin; ka dha; da ka; dha dhin; dha ga ta ..."[22] We can only guess, then, whose triumph these talking rhythms are intended to accompany.

Interestingly, in an interview in *New Musical Express* following the release of the album in 1982, Kate Bush stated: "For me the singer is an expression of the song—an image should be created for each song, or at least each record" (Cook 1982). In other words, rather than being true to a particular preestablished artistic persona, Bush regards both singing style and vocal production as tools that should be used in the service of the given song. This approach emerges as different from the classic-rock approach to vocal production, where the vocal—for a fascinating mixture of commercial and ideological reasons—is used as a sonic trademark, and where any manipulation of the lead vocal away from the singer's "true" (inner) expression is regarded as "falsifying" the voice of the singing subject. In contrast, Bush's approach to vocal production is theatrical or, one might even say, postmodern, in that she obviously stages the singing subject and even splits it into different characters. In fact, this "theatrical" approach to vocal production became widespread and typical for a wave of explicitly staged artist personas in the years to come, encompassing, for example, highly successful artists such as Madonna and Prince. As part of the mainstream, the new digital signature represented by the surrealistic virtual spaces produced by digital delay and reverb was used to realize various postmodern aesthetic visions through its loosening of the bond between the personal (or "real") life of the artist and his or her artistic persona and vocal performance.[23] In "Get Out of My House," as well as several other tracks on *The Dreaming*, Kate Bush actively used the new digital possibilities of fabricating virtual spatialities to support and enhance the different characters and emotions, as well as the juxtaposition of her different vocal personas. In "Get Out of My House" in particular, the theatrical vocal performances, gruesome undertones of the lyrics and general eerie atmosphere produced by the processing effects of digital reverb and delay combine to generate a musical meaning that is more than the sum of its parts, a meaning that is, appropriately, quite horrible.

Surreal or Naturalized?

As we have seen, digital reverb and delay effects are able to generate aesthetic meanings of their own as well as underpin the meaning communicated to the listener by other musical aspects, such as lyrics and vocal performance. The work of Kate Bush and her production team to exploit the possibilities of these then-new technologies made "Get Out of My House" the track it is. While some of their effects could have been achieved with analog technology, digital-era possibilities simplified, encouraged, and ultimately standardized these effects in a remarkably short time, as this track clearly demonstrates. The fabrication of new virtual spatialities became a typical trait of music production in the early 1980s and, as such, a signature of the penetration of new digital tools into the field of popular music.

In his thorough discussions of how musical spatiality has been fabricated in popular music recordings during the period between 1900 and 1960, Peter Doyle points out that, despite the fact that reverb and echo effects are ubiquitous in contemporary popular music, "questions of how these sonic variables might bring about an affective outcome in listeners have gone largely unasked" (Doyle 2005, 5). We have addressed two such questions in this chapter, namely, how the processing effects of digital delay and digital reverb have been used as compositional tools in music production, and in what ways these effects are experienced by the listener. Concerning the second question, it is important to emphasize that the ability of delay and reverb to create virtual spatial environments is premised on the listener's tendency to compare the musical sounds with those produced by actual physical spaces. One possible consequence of this tendency is an unbalancing sense of surreality in the face of effects that are unlikely, to say the least, in the real world. For example, the digital delay in "Get Out of My House" clearly differs from any naturally occurring delay, as described above, and the same goes for the gated reverb, reverse reverb, and general spatial collaging used in the song. This may be one of the reasons why several journalists described "Get Out of My House" as bizarre, odd, weird, and surrealistic.[24] The technological mediation in "Get Out of My House" draws attention to itself and comes forward as quite opaque as the split between its sounds and their putative spatial origins is made obvious.

Although the spatiality of "Get Out of My House" was almost certainly experienced as surreal upon its release, sonic designs such as Bush's generally tend to become relatively naturalized with the passage of time. The reason for this is that our ears "tune to" or adjust to new sonic environments with dispatch. As we adjust to new musical expressions brought about or inspired by new technologies, we come to hear them as appropriate—even "natural"—and we will then judge the next round of innovations against them. Any given experience with a musical environment promptly becomes a reference point as we structure and comprehend the next environment. Or, as Smalley points out, "The perspective of the acousmatic image has evolved its own conventions" (Smalley 2007, 50). If we compare the sounds in "Get Out of My House" to sounds from actual spaces, the music will certainly appear surreal. If we compare it to other tracks like it, especially from the early to mid-1980s, it will sound less remarkable (though certainly virtuosic nevertheless). The experience of musical spatiality as natural or surreal thus depends on the frame of reference within which we meet its affordances. Instead of pointing to an inherent quality of the phenomenon in question, then, the impression of something as surreal or natural tells us something about what we are comparing it to.

While Kate Bush's music likely shocked the early 1980s sensibility regarding a "natural" sonic environment, today we have become very comfortable with the juxtaposition of different sonic spaces in music. Even so, the music's surreality might persist for us now. Our comparison of the virtual sonic environment of a recording *both* to actual spatial environments and to contemporary naturalized musical environments presumably generates the richness of our experience with music like this. Even though we know that anything goes here, we still perceive the eeriness of the juxtaposition of different vocal personas in Kate Bush's music and enjoy the way opaque mediation flouts consequence by evoking our familiarity with real physical spaces even as it subverts it.

3   The Instrument Formerly Known as the Machine: Hyperaccuracy and Sonic Richness in Prince's "Kiss"

In the early 1980s, when new digital reverbs, synthesizers, and samplers were introduced to the field of popular music, producers were still trying to avoid noise, and imprecise timing, on records. It is therefore not surprising that some of the most remarkable features of the new digital tools involved ways in which high fidelity in sound, as well as temporal precision, could be taken to a new level. Even though sequenced "machine" rhythms and synthesized sound had been part of popular music for more than a decade, the digital contrast to the previous analog modes of production was striking. This new hyperaccuracy in sound and timing made the fuzzy, distorted product of analog synthesizers, as well as the skeleton-like rhythmic patterns and artificial-sounding "drums" of previous drum machines, seem completely obsolete.

In addition to addressing old problems regarding noise and imprecision, the sounds and procedures made possible by digital technology presented music producers with a new palette of compositional materials and tools. This palette is closely connected to what has survived as the signature "digital" sound of the 1980s: machinelike grooves featuring an unprecedented richness and clarity of sound. This signature digital sound arose from a combination of several factors. Digital synthesizers and samplers offered a new richness *of sounds* through their presets and sample libraries, as well as a new richness *in sound*, since the synthesized and sample-generated sounds of the new digital tools were significantly more complex than those produced with previous analog equipment. The noiselessness of digital equipment also enabled a new sonic richness owing to the presence and clarity of the upper frequencies as a consequence of the absence of noise and distortion. As discussed in the previous chapter, this could first be heard in the use of digital reverb, which had a fascinating "sparkling" effect. However, the new richness in sound was also a consequence of digital sampling. The digital sampler could digitally record (sample) and reproduce short sounds, which could be applied as sound sources for percussive instruments in sample-based drum machines or for new, natural-sounding presets in keyboard-based instruments. The sampler thus made it possible to mimic complex acoustic instruments in a new and convincing way. Finally, the MIDI protocol became the standard for communication between digital instruments. This meant that digital music instruments could be connected to and controlled by each other, even when manufactured by different companies. The timing profile of one MIDI track, for example, could be used to control another track with a different sound source (this is one of the prominent elements of the sound of "Kiss"). MIDI also took timing precision to a new level in the sense of placing musical events on a metric grid, because all of the tracks in a production, not only those produced by the drum machine but also those produced by synthesizers and samplers, could now be coupled and aligned to the same grid through quantization.

In the following we begin with a presentation of these technological developments, and then turn to how Prince took advantage of them in the song titled "Kiss," which is a striking example of both signature 1980s digital sound and Prince's very special take on it.

A New Palette of Sounds

In analog sound synthesis, the sound source is one or more voltage-controlled oscillators producing periodic activity in the audible range. The sound of the oscillator can display different waveforms (for example, sawtooth or square-pulse) or generate noise. One modifies this signal by patching it through other components, such as filters, modulation generators (for example, a low-frequency oscillator or LFO), envelope generators (EGs), or amplifiers. Parts of the signal can be removed using the "hi-pass" and "lo-pass" filters; vibrato can be added, either to the filter or the oscillator or the amplifier, by way of the LFO; various temporal envelopes (attack, decay, sustain, release) can be applied to the process via other components; and so on.

Digital sound synthesis, on the other hand, is the process of generating a stream of numbers virtually representing points in the curve of an audio waveform. In contrast to sampling synthesis (see below), these numbers are *not* picked from an existing waveform but constructed from scratch, so to speak. The sound can be heard only after these numbers have gone through a digital-to-analog (DA) converter, which converts the numbers to a continuously varying voltage that can in turn control the vibration of a loudspeaker (see also chapter 1). The first experiments with computer-based digital sound synthesis took place at Bell Laboratories in the 1950s. Producing one second of continuous sound required tens of thousands of numbers. At this time, of course, computer power was very limited, and the calculation of these sound-producing numbers was thus so demanding that it had to be carried out in New York by the American computer corporation IBM (International Business Machines). Back at Bell Lab, chief engineer Max Mathews and his colleagues then converted the numbers on the magnetic tape to audible form (Roads 1996, 87).[1]

An important step in the development of the digital synthesizer was the introduction of unit generators—signal-processing modules with functions resembling those found in analog synthesizers, including oscillators, filters, and amplifiers. These could be combined in different ways to form patches that produced a variety of synthesis algorithms of considerable complexity, which in turn resulted in richer and more complex sounds.[2] Digital sound synthesis is based on fixed-waveform synthesis, which means that the computer repeats a short waveform that has been programmed in advance. Making musical sounds, in this case, means constructing algorithms to be executed at a later stage by the computer. Such non-real-time software synthesis allows for extremely detailed and complicated procedures, but, as Curtis Roads points out, there are obvious disadvantages as well. One has to wait for samples to be computed, and sound is disconnected from real-time human gestures. The sound cannot be shaped as we hear it being generated (in real time), and, according to Roads, "The stilted quality of some computer music derives from this predicament" (1996, 105).

Digital sound synthesis did not influence the sound of popular music to any great extent until the development of commercial digital synthesizers in the form of keyboard instruments. The first influential commercial digital synthesizer was the Yamaha DX7, launched in 1983 (Holmes 2012, 346). Tone generation with the DX7 was based on frequency modulation (FM) synthesis, in which a modulator frequency modulates a carrier frequency so as to produce bands on either side of the carrier frequency. By increasing the amplitude (of the modulator), one produces a richer spectrum of these side bands (Manning 2004, 193–194)—this feature is typical for the sound of many instruments, including strings, horns, and drums (Rossing, Moore, and Wheeler 2002). Analog FM synthesis had been in use for radio transmission (FM broadcasting) since the late 1930s, and in the early 1970s, John Chowning, a composer and percussionist working at Stanford University, developed a frequency modulation *algorithm*—that is, a digital version of FM synthesis—that proved especially useful for synthesizing sounds with a rich and time-variable spectrum of overtones. He licensed his algorithm to Yamaha in 1973 for use in the DX7. The DX7 thus had a much richer and brighter sound than its analog predecessors, which made it very popular in the 1980s. It was capable of sixteen-note polyphony and featured a total of thirty-two algorithms (each one a different arrangement of its six sine-wave operators), allowing for a great deal of programming flexibility. However, the programming process was found by most to be quite complex, and the instrument is now primarily associated with its characteristic preprogrammed presets, such as electric piano, bells, and other "struck" and "plucked" sounds with complex attack transients (the first, unstable phase of the sound that often gives the instrument its particular sonic character, or timbre). In addition to offering a new richness in sound, the DX7 thus also represented a new richness of sounds through its presets, a palette of new, more organic-sounding synthesized sounds. In Prince's "Kiss," for example, the melodic keyboard hook entering the song in the second verse is the unmistakable sound of the marimba preset of a DX7 (Brown 2011, 111).

In short, the DX7 and other digital synthesizers became extremely relevant to the characteristic warm and rich synthesized sound of 1980s pop songs owing to both their noiseless richness in sound and the new richness of sounds provided by their presets. However, whereas the digital sound synthesis used in the new synthesizers relied on constructing sound from scratch, another form of digital sound synthesis would become even more relevant to computer-based *grooves*—namely, sampling synthesis, or the process through which an existing sound is recorded and digitized for use in sample-based digital instruments. This development, together with the standardization of a code for communication among digital music instruments (the so-called MIDI protocol), was crucial to the digital signature of Prince's "Kiss," for example. This song fundamentally relies on two innovations: the possibility of using natural drum sounds in a sequencer groove (enabled by sampling synthesis), and the orchestrated coupling of multiple digital devices (enabled by the MIDI protocol).

Sampling and Sample-Based Instruments

Sampling, in the sense of using small fragments of recorded sound as musical building blocks,[3] is not in itself a new practice but rather dates back to the 1920s in the West. Pierre Schaeffer's *musique concrète*, for example, is based on a form of analog sampling via microphone and magnetic tape. Rather than using sound synthesis to generate sound for their compositions, composers in this milieu recorded sound and then worked directly with these "concrete sound objects" (Schaeffer 2004). There are also pre-digital instruments based on sampling. In the field of popular music, the Mellotron of the late 1960s and 1970s is probably the best-known example. This instrument contained a number of rotating tape loops and was used by numerous bands, especially within the progressive and symphonic rock genres, to create "orchestral" or "choral" sounds.[4]

Whereas these examples of analog sampling are based on magnetic tape recording technology, digital sampling, again, involves creating a numerical representation of the audio waveform (for an explanation of the principles behind digital sampling and recording, see chapter 1). The initial capacity of digital electronics to record and store sound in digital memory chips was first exploited via digital delays (see chapter 2). The first digital sampling devices, then, were recording-studio delay units designed to "enrich the sound by mixing it with a sampled version of itself delayed by several milliseconds" (Roads 1996, 120). Toward the end of the 1970s, when computer memory became cheaper, digital sampling began to be used in instruments as well.[5] The first commercial sample-based keyboard instrument, the Fairlight CMI Series 1 (Fairlight Inc.), was introduced to the market in 1979. The Fairlight CMI Series 2, which offered a better sampling rate and thus better sound quality, arrived in 1982, at a cost of approximately $35,000 (Creative 2015; Leete 1999). Prince, along with Kate Bush and many other influential musicians and producers, bought one in the early 1980s (Leete 1999; Brown 2011, 109). A cheaper alternative, the Emulator (E-MU), was introduced in 1981 with a sampling time of two seconds at a third of the price ($10,000) of the Series 2 (Creative 2015), and a wave of increasingly powerful and inexpensive sample-based keyboard instruments soon followed (Holmes 2012, 492).

These tools were all designed to easily and efficiently imitate both acoustic and electronic instruments using prerecorded samples or samples that were recorded by the sampler itself. When one wanted to evoke an instrument with sustain, such as the piano or organ, the audio waveform had to be looped in its sustain phase, so that the sound would last as long as the key was pressed down. When one wanted to synthesize sounds with pitch, the sounds had to be pitch-shifted to suit the different keys of the keyboard (the early sample-based instruments didn't have enough memory capacity to assign one sample to each key). Despite the new richness of the spectral envelope, thanks to the completely noiseless replication and manipulation of sounds from real instruments, the resulting music often sounded somewhat "mechanistic" owing to the lack of any development in the sounds themselves: because of the looping of the sustain phase of the sounds and the uniformity of tone (timbre) among the different keys (the tones sounded the same despite being played in different parts of the register), the temporal envelope was experienced as quite poor compared to traditional instruments played by musicians. Thus, acoustic instruments, such as brass or strings, that were generated using sampling synthesis sounded "dead" and often required artists or producers to overdub the sampler with an actual acoustic instrument—a real trumpet or violin, for example, might double the synthesized lead part to supply natural "life" to the sound.

Short percussive sounds, on the other hand, could be recorded and played back in their entirety. In addition to sample-based synthesizers, then, an important area of application for digital sampling technology in the commercial market was drum machines. The drum machine does not in itself require digital technology, of course—the first drum machine for musical use, the Rhythmicon, was actually introduced in 1930 by Leon Theremin (Theremin 2014).[6] Furthermore, analog drum machines could be programmed like their digital heirs. The key difference between analog and early digital drum machines was that the former used analog sound synthesis, rather than samples, for their drum sounds. For example, in an analog drum machine, a snare drum sound typically came about via subtractive sound synthesis, using a burst of white noise as its starting point. This meant that the final sound was not particularly close to the real instrument. On the other hand, some of these sounds were very interesting indeed, and each model tended to have a unique character. For this reason, many analog drum machines lent themselves to very creative applications, and some of them have since achieved cult status—notably the Roland TR-808 and TR-909 [Zeiner-Henriksen 2010a, 2010b].

In contrast, the digital drum machine used samples of real drums as its sound source. Unlike more sustained sounds, such durations presented few challenges to the sampling process, and percussive sounds typically lacked specific pitch as well. Because percussive sounds have very complex and disharmonic spectral features and are almost impossible to produce in a realistic way via other forms of synthesis, sampling also meant, in this case, a great leap forward in sound quality. The new sampling technology thus suited the reproduction of percussive instruments very well.

The very early LM-1 from Linn Drums, like most other digital equipment at the time, was very expensive but played a huge part in popular music production nevertheless during the first half of the 1980s. Only about five hundred of them were ever made, but as with the Fairlight CMI instruments, the list of LM-1 owners is impressive and includes Peter Gabriel, Stevie Wonder, and Prince, who used it on nearly all of his most popular recordings, including *1999* and *Purple Rain* (Buskin 2013). It consisted of a programmable digital sequencer that triggered recordings of real drums according to the programmed pattern, and its sound became definitive for 1980s "digital" pop. Many of its drum sounds were in fact composed of two chips that were triggered at the same time, and each sound was individually tunable.[7] The console also included a built-in thirteen-channel mixer (one channel for each sound), as well as individual output jacks, which enabled its integration with existing recording equipment in a way that had never before been possible for a drum machine.

A cheaper version of the LM-1 called the LinnDrum (or the LM-2) was released in 1982 at the price of $2,995 (Delton 2012b). This model included five external trigger inputs, but the possibilities for tuning the drums were reduced. A later model, the Linn 9000, which was released in 1984, was used on Prince's "Kiss." By then the MIDI code had become an industry standard, and the Linn 9000 was fully equipped for MIDI communication, which meant that its drum sounds could be controlled by an external device, and vice versa—the sequencer of the Linn 9000 could control sounds in other digital modules.

Standardizing Communication: The MIDI Code
Following the success of the LM-1, other manufacturers also began to produce digital drum machines.[8] These early models were based on different digital standards, and, as was the case with synthesizers, compatibility

among machines remained a thorn in the side of manufacturers. The multifarious nature of synthesizer design meant that each manufacturer had been defining data in its own way. Eventually, some developed digital interfaces that would allow one to link multiple Korg, or Roland, or Yamaha synths, but no common code for communication between digital instruments existed.[9]

Manufacturers began to worry that this lack of compatibility would inhibit widespread use of digital equipment and ultimately hurt sales, and there was soon talk of a "universal" digital communication system. In 1981, Dave Smith and Chet Wood from Sequential Circuits presented a paper to the Audio Engineering Society that described a concept for a Universal Synthesizer Interface using regular quarter-inch phone jacks. At the following NAMM (National Association of Music Merchants) show in January 1982, a meeting took place between the leading US and Japanese synthesizer manufacturers that produced certain improvements to the interface, which was then known as MIDI—an acronym for "musical instrument digital interface." In December 1982, MIDI appeared on its first instrument, the Sequential Prophet 600. Roland's JP6 followed shortly thereafter, and the two were successfully connected at the NAMM show in January of the following year.

In 1983 the MIDI specification was all of eight pages long and defined only the most basic instructions, such as the choice of channel and how to control the output volume. Since 1983 the protocol has grown to encompass additional concepts including standardized MIDI song files (General MIDI, 1991) and new connection mechanisms such as USB, FireWire, and Wi-Fi. However, the way MIDI works has not in fact changed since its arrival, and the industry agreement to adopt a standard and royalty-free technology such as MIDI remains, in this context, a singular achievement. MIDI allowed for entirely new ways of using old tools, including synthesizers, sequencers, samplers, and drum machines. MIDI also enabled computers to be applied to the music-making process in a new way and thus sparked wide-reaching interest in the general integration of the various tools that were required for record production. According to Richard J. Burgess, MIDI was a transformative technology: "MIDI increased access to a rapidly expanding palette of sounds and began the convergence and integration of the various technologies and methodologies. The many all-in-one units and subsequent software programs, along with falling prices, signaled the beginning of the democratization of the production process" (Burgess 2014, 142). Conversely, the capacity of these new units was limited: "It was still not practical to record full-length vocals or other acoustic parts, the professional studio was still a necessary, if now postponed and shorter, step" (ibid.). This early effort, of course, culminated in the introduction of digital audio workstations (DAWs) in the early 2000s, which, combined with the ongoing increase in the power of personal computers, has brought about a revolution of sorts in the recording industry.

MIDI increased the richness of available sounds because of the ways in which it allowed one to combine sounds from different digital sources. One set of MIDI signals could be used to control many instruments at the same time, which made it easy to dub or duplicate a pattern across instruments. Put differently, the "event list" of one MIDI track could be used to control the output of another track with a different sound source. As we shall see, this exact operation was done on "Kiss," thanks to the pioneering capacities of the Linn 9000 in this regard.

MIDI also led to a new level of precision in timing by way of quantization. Quantization is the process of automatically positioning performed musical notes according to an underlying predetermined temporal grid that commonly represents beats and subdivisions of beats. In the early stages of MIDI, quantization was either on or off, meaning that there was no flexibility in the process. When fully quantized, all of the onsets and offsets of the musical events in a song, including those with only minor "expressive" deviations, were moved to the closest position in the predetermined grid. This resulted in a sonically distinctive hyperaccuracy of timing that became associated with the machinelike musical expressions of the 1980s. This precision was possible to achieve to a significant extent with analog drum machines as well,[10] but the MIDI sequencer allowed one to apply quantization beyond the drum parts of the groove. A pattern of MIDI events could be programmed either visually by entering MIDI events on a screen, or physically by playing the pattern and quantizing it so as to control any sound source that was amenable to the MIDI protocol. This combination of MIDI and the new sounds produced through digital sound synthesis or digital sampling synthesis produced an unprecedented hyperaccuracy in the temporal domain that governed the whole groove, not only the drums. It also brought about a new sonic clarity or "realism" in the sound. Together, these qualities would define the digital groove of the 1980s.

The Digital Groove

The song "Kiss" was originally intended for a band called Mazarati, which was signed to Prince's new label to record an album produced by Prince's sound engineer and collaborator David Z (born Rivkin). The band needed a single and received a demo of "Kiss" from Prince. When David Z returned to the studio the day after finishing what he thought was going to be a Mazarati single, he found that Prince had replaced the chesty vocal of Mazarati's Tony Christian with his own falsetto singing, removed the bass line, and added the characteristic guitar riff. Prince had been so thrilled with the result of David Z's efforts with the band the previous day that he had decided to take his song back (Nilsen 2004, 219–220).

The demo of "Kiss" consisted of Prince singing one verse and one chorus an octave below where the song ended up in the recorded version, while accompanying himself on the acoustic guitar. Today, the demo comes across as the sketch of a straightforward pop tune built on a twelve-bar blues scheme. In an interview in *Sound on Sound* (2013), David Z recalls that he did not know what to do with the demo—to him, at first, "Kiss" sounded like a "folk song." However, as he was heavily into MIDI-compatible samplers, synthesizers, and drum machines, he promptly programmed its beat on a Linn 9000 digital drum machine (see above), which featured a thirty-two-track MIDI sequencer, eighteen built-in sampled drum sounds connected to velocity-sensitive drum pads, a mixer section, a programmable hi-hat decay, and an LCD screen.[11] Because the Linn 9000 included a MIDI sequencer, it could be connected to and control other sound sources, such as a synthesizer or a sampler.

A key feature in the "Kiss" groove is the hi-hat track, and especially the ways in which its peculiar rhythm is allowed to control the entire groove. After programming the basic rhythmic pattern, David Z began experimenting: "I ran it [the hi-hat track] through a delay unit and switched between input, output and in the middle [a blend of the two]. That created a very funky rhythm" (Buskin 2013).[12] When the hi-hat track was switched to output, then, it had a delay on it, and when it was switched to input, it did not. Afterward, David Z played the open chords from Prince's demo on an acoustic guitar. He strummed the first beat of each bar (Dr. Fink, in Brown 2011, 111), recorded it, and then had the hi-hat track trigger the sound of the guitar track by way of a noise gate.[13] The input of the gate was controlled by the hi-hat track, so that whenever there was sound above a certain level from the hi-hat, the gate opened and the sound of the recorded guitar "came out." When the sound on the hi-hat track fell beneath that level, on the other hand, the gate closed and the guitar was silenced. As a consequence, the guitar no longer sounded like a guitar—one has the impression of chords and chord changes of an uncertain instrumental origin, pulsing along with the hi-hat pattern. According to David Z, "The result was a really unique rhythm that was unbelievably funky but also impossible to actually play" (David Z, quoted in Buskin 2013).

The groove of "Kiss" is quite open, with kick drum strikes on beats one and two-and alternating with snare strokes on beats two and four (see figure 3.1). The hi-hat sounds an offbeat-oriented pattern of sixteenths, some of which are actually entries in the sequencer program and some of which are probably produced by the delay. The pattern of the acoustic guitar sound, which, in its gated form, sounds more like a synthesizer (probably because of the absence of the transients of the acoustic guitar), follows the hi-hat pattern except for the last two strokes, which consist of hi-hat only. As we can see in the spectogram in figure 3.2, the timbre and intensity of the hi-hat/acoustic guitar pattern vary constantly, producing a distinct microrhythm that contributes to the characteristically machinelike yet nevertheless dynamic feel of this groove. This aspect of the microrhythm of the groove cannot be captured by notation, because it derives from sound-related aspects of the hi-hat strokes rather than their temporal placement and timing.

Contrary to what we might expect, given that the groove was programmed in 1985, the timing of the different rhythmic events is partly irregular—not all of the hi-hat strokes, that is, are completely on the grid of the sequencer. Again, this is probably because the delay unit produces some of the "echoing" hi-hat strokes. Though the "programmed" feel of the groove is unmistakable—it belongs to the machine—its combination of accuracy in timing and very dynamic use of sound is remarkable. First of all, the obviously programmed snare drum (see its alignment versus the grid in figure 3.1) features a quite rich and realistic sound that is probably one of the standard samples on the Linn 9000 drum machine. The kick drum sound is also unusually rich for a programmed groove, but, contrary to the snare, it does not sound realistic. Instead, the kick drum, which was placed on a separate track in the production process, was heavily processed

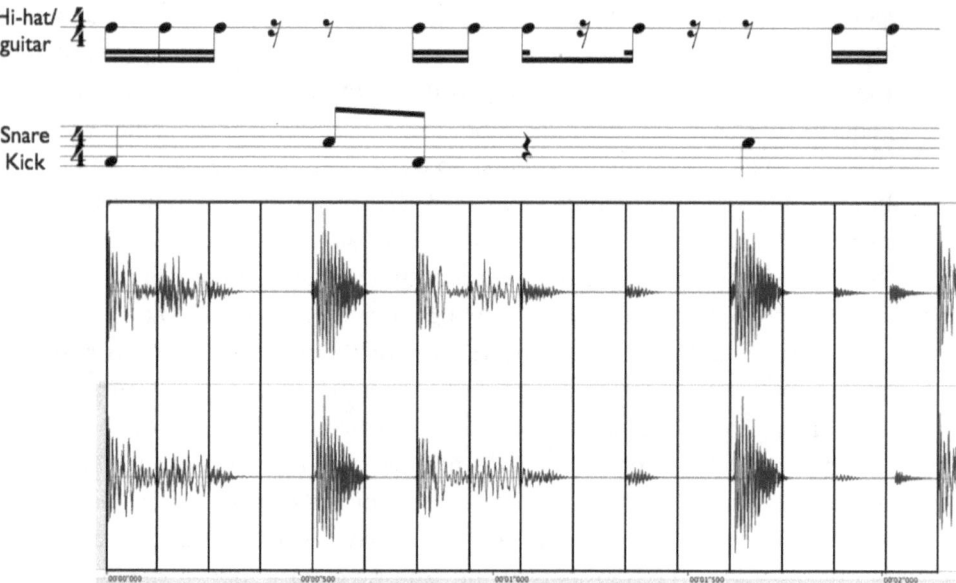

Figure 3.1
Transcription and waveform (amplitude/time) of the basic groove in "Kiss." Grid on eighth notes.

in order to encompass the role of bass guitar as well: "As for the lack of bass guitar, we always ran the kick drum through an [AMS] RMX 16 and put it on the Reverse 2 setting to extend the tail of the reverb. That served as a kick drum and a bass, and it was a signature sound that we used all the time with Prince. We didn't need a real bass" (David Z, quoted in Buskin 2013). The Neve AMS RMX-16 was a high-end digital delay and reverb unit that offered presets imitating real acoustic spaces, as well as certain surreal effects.[14] The preset labeled "reverse 2," for example, which was one of its special effects programs, produced the reverse of natural reverberation (AMS Neve 2000). Depending on the setting of the decay control, the reverberating sound built up for a period of time and then suddenly stopped (see chapter 2 for further description of reverse reverb). This was used in "Kiss" to augment the drum sound on the first beat of the bar (Daley 2001), the result of which can be seen in the spectogram above (see figure 3.2). As a consequence, in addition to fulfilling the roles of both kick drum and bass guitar (or "base," as David Z recalls they dubbed it [Brown 2011, 110]), the kick drum becomes a little groove in itself: instead of merely articulating

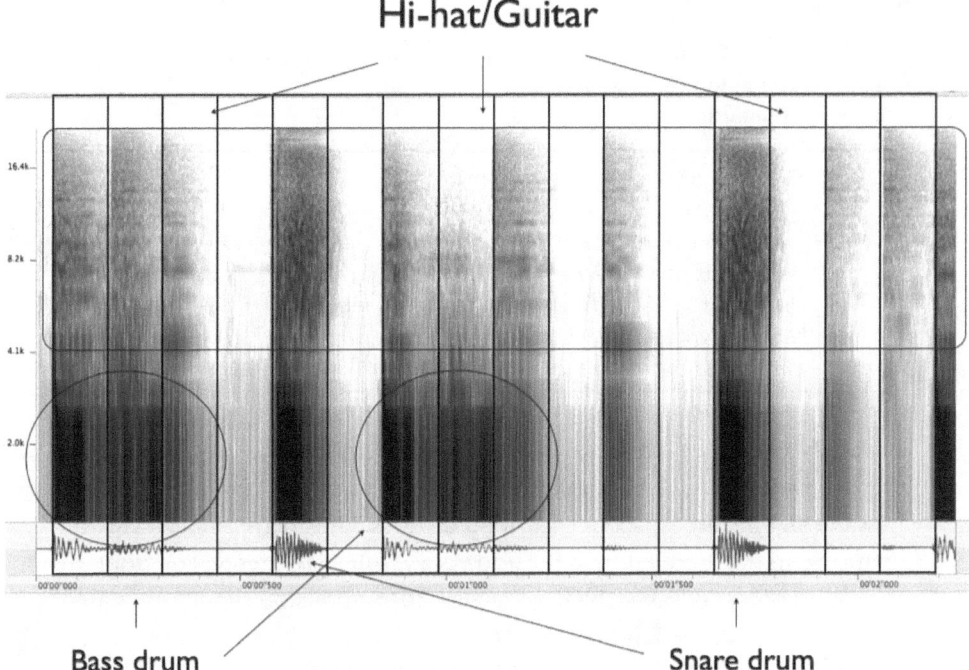

Figure 3.2
Spectrogram (0–20,000 Hz) of the basic groove (bar 1) in "Kiss" (Amadeus Pro 1.5.4). Grid on sixteenth notes. Reverse reverb on bass drum indicated by circles.

the beat as a point in time, the drum sound is extended and given a specific dynamic and timbral shape that develops over time.[15]

In addition to its remarkable groove, "Kiss" stands out for its minimalist approach. Prince and David Z used only nine or ten tracks for the song (see table 3.1),[16] and mixing it took less than five minutes, according to David Z. A small amount of tape delay was added to the guitar track, but otherwise the mix was just the result of "Prince pulling back and turning off faders" (Buskin 2013). David Z also recalls that the starkness of the mix actually made him a little uneasy: "I reached over and snuck a little bit of the piano [DX7] back in" (Daley 2001).

The machinelike character of the groove is emphasized by the fact that there are no "small sounds" or fillers in it. The basic pattern is repeated throughout the song, without variation, which foregrounds the fact that it is produced by a machine (a sequencer). In addition to the sparse use of instruments and the absence of fillers in the sound, the production is

Table 3.1
The tracks of "Kiss"

1	kick drum (with reversed reverb)
2	other drums
3	hi-hat
4	acoustic guitar triggered by hi-hat
5	guitar hook
6	wah-wah guitar (with a little tape delay)
7	keyboards (marimba preset on DX7)
8	backing vocals L (Mazarati original backing vocals)
9	backing vocals R
10	lead vocal (recorded using a Sennheiser 441 microphone)

strikingly dry, especially if we take into consideration the generous use of digital reverb in mainstream pop in the first half of the 1980s.[17] This dryness is a significant aspect of the sound of "Kiss." It is perhaps most clearly expressed in the production of the lead vocal, which is extremely up front in the sound box and leaves the listener with the impression that it originates almost outside of the loudspeaker. According to David Z, the lead vocal was left completely dry, which was—and is—absolutely unconventional.[18]

Another significant aspect of the extreme proximity of the voice is its clarity and presence at the high end of the frequency spectrum, which was, according to David Z, achieved using a microphone that boosted the frequencies around 3 kHz (a Sennheiser 441 with a bass roll-off switch [David Z, in Daley 2001]). These qualities are also related to the absence of noise and distortion in the production as a whole, owing to the use of digital equipment. The original backing vocals of Mazarati, which were kept in Prince's version, are also very present in the sound. The sound of the backing vocals is very different from the lead vocal, but they convey the same feeling of a very articulate foreground and no background. Even though they are sung forcefully with a chesty voice, they have no depth or audible reverb.[19]

The low level of background noise is an important prerequisite for the experience of hyperpresence in the sound of "Kiss," because it forces us to reckon with the particular sonic character of the song, as well as the minute dynamic details of the musical elements in the foreground. Again, this is most striking with regard to the lead vocal—a digital nothingness, that is, underlines the extreme proximity of Prince's vocal delivery, and we hear every little breath, sigh, and, not least, smack of Prince's lips. But the machinelike groove attracts our attention as well. And thanks to the groove's ultimately quite subtle dynamics—the interesting "inner life" it achieves through precisely those new possibilities provided by the digital tools of the era—it thrives on this extremely exposed positioning in its sound.

Sophisticating the Sounds—Computerizing the Groove

Repetitive programmed grooves were nothing new, even in Prince's era, and rich, organic sounds were, of course, old, but the *combination* of the two was unique to MIDI and digital sampling, and it opened up opportunities for controlling grooves and processing superior sound in new ways. Sometimes this new technology was used to solve old problems, as was to a large extent the case with Michael Jackson's *Thriller* (Epic, 1982), which comes forward as an eminent example of the advances in sonic clarity and temporal precision that digital technology brought about. In the case of "Kiss," however, MIDI and digital sampling brought about something completely new—the strikingly dry, minimalist sound made "Kiss" into a model for state-of-the-art pop production for the rest of the decade. The song demonstrated the need for getting the machine to groove, in the sense of producing interesting sounds at the microlevel. Prince and David Z achieved this by tweaking the machine to its limits to create compelling new dynamics within a single prominent sound and within the period of time that constituted the basic repetitive pattern of its groove.

Prince's abrupt turn toward a minimalist synthetic groove aesthetics in "Kiss" was also truly remarkable given the guitar- and rock-oriented sound of *1999* and *Purple Rain*, with which he conquered the mainstream pop/rock audience. However, from a different perspective it represented a return of sorts to his premainstream productions of the late 1970s. Prince belonged to a musical scene in Minneapolis where machinelike funk, played on synthesizers, was the rule rather than the exception (see, for example, songs like "Soft and Wet" [1978] and "Controversy" [1981]). The new possibilities provided by MIDI and digital sequencing in the early 1980s thus fitted very well with the repetitive, groove-based, synthesizer-driven musical idiom of Prince *before* his crossover success. Nonetheless, the radical use of technology was crucial to the leap in Prince's career that "Kiss" represented. The minimalist richness of the groove of "Kiss" attested to a very creative use of contemporaneous digital tools, particularly the digital drum machine with sampled drum sounds and the possibility of connecting and controlling different digital devices by way of the MIDI protocol. Thus, instead of being associated with "authentic" black musical roots, "Kiss" comes forward as an insistently modern and "color-free" manifestation of funk. The state-of-the-art use of digital technology clearly contributes to this touch of the avant-garde.

As discussed above, grooves made by sequencers in the early 1980s were often perceived to be nonhuman and "mechanistic" (they were, of course, also often intended to sound this way). Their mechanistic feel can be attributed to the fact that there was no flexibility in the temporal placement of rhythmic events, which were all forced to take place according to the grid of the sequencer. Rhythmic patterns consisting of grid-ordered events seem to lack a human touch—that is, the deliberate and unintended variations that musicians add to their performances. However, this lack of flexibility in timing is only partly responsible for the machinelike or "stiff" impression of sequenced grooves. As demonstrated by Zeiner-Henriksen in his analysis of electronic dance music from the 1970s through the 1990s, onsets of events in this genre are usually on the grid as well, but the ways in which the sounds of the rhythmic events are shaped make the groove experience less stiff and even induce motion in the listener.[20] An equally important aspect of machinelike musical rhythm, then, is the absence of the small variations in *intensity* and *timbre* that are always present in a played series of drum strokes. In played music, the force with which drums are struck will not be constant. There will be deliberate and predictable variations in the *sound* (in addition to timing variations), depending on the style and structural context, and there will also be some unavoidable small-scale, random variation (no matter how much the drummer tries to minimize it). Conversely, a drum machine's strokes all sound exactly the same, for better or worse.[21]

A fascinating and important aspect of "Kiss" is how Prince and David Z managed to humanize the machine feel that was commonly associated with the use of drum machines at the time, thus challenging the discursive dichotomy of human versus machine. This dichotomy was still very present in 1980s popular music. On the one hand, there were organic-sounding, "authentic" styles, such as rock, country, and R&B, which claimed no (or at least no audible) exploitation of digital technology at that time; on the other hand, many artists produced entirely digital dance music in the aftermath of Kraftwerk's futuristic admiration of machine aesthetics and the computer as aesthetic object, as expressed in their albums *Man-Machine* (1978) and *Computer World* (1981). The increase in sonic richness in "Kiss" (owing to the use of sampled actual drum sounds), and the introduction of small variations in the sound of hi-hat strokes (through manipulation of the digital delay and the MIDI-enabled connections among different devices, so that the rhythm of the hi-hat track could "merge" with the sound of the acoustic guitar) challenged the traditional conception of the machine as something stiff or fixed.

As we have seen, the new "humanized" feel of "Kiss" was a product of new *technology*: thanks to the combination of MIDI and digital instruments, the sound of one instrument and the rhythm of another could be merged in a way that humans clearly cannot achieve. This distinctive hyperaccuracy in timing was reminiscent of former machinelike expressions, but the new richness and clarity in sound characteristic of the digital groove exceeded the characteristics of the machine as we used to know it. In this respect, "Kiss" was only the beginning: in the years to follow, the digital instruments once thought of simply as machines would occasion the utter transformation of popular music from "either/or" to "all of the above," as both performers and their equipment improved and it became increasingly difficult to distinguish between human- and machine-made musical utterances.

4 The Rebirth of Silence in the Company of Noise: Portishead Going Retro

With digital technology, silence was reborn. While analog media presented us with silence of a sort, at least when compared to the sound quality of electromechanical media, digital silence made obvious all of the noise that we had previously ignored. Before the introduction of digital technology, recorded sounds had always been enmeshed in the noises inherent to the mediating process. Digitally recorded sounds, on the other hand, had no noises accompanying them, and consequently, digital silence took us to a "musical" place we had never been before. Reducing the distraction of background noise and improving sound quality have always been motivating factors behind new developments in sound reproduction technology, and with its total silence, digital recording thus represented a dramatic change in this narrative of audio "progress."

While digital recording technology's fidelity to a sound's source seemed perfect, the digital medium was not universally regarded as the perfect medium for recording music. In a countercultural reaction to the musical embrace of high-fidelity sounds, several musicians went lo-fi, exploring the unique sonic signatures of various predigital recording and playback media, instruments, processing effects, and other musical equipment. As Joseph Auner (2000) and Stan Link (2001) point out, the practice of exploring and reacquiring noises and lo-fi sounds from predigital eras was very common in the 1990s. Still, this trend implied not necessarily a complete rejection of the digital but rather a counterpoint to it. As I will discuss in this chapter, the digital signature was sometimes even explored and exposed on its own terms, silent or otherwise. And various combinations and interminglings abounded: the old sounds were often put into a digital frame, for example, allowing us to hear them anew. By providing a new basis for comparison, the silence of the digital in effect reintroduced analog's various forms of noise.

In this chapter, we will first situate the digital signature of total silence in a historical lineage, representing it as a radical departure from earlier recording and playback technologies. Next we will point to how digital silence encouraged the aforementioned schizophonic tendency to revisit media signatures from the past. We will then analyze "Strangers" (*Dummy*, Go! Discs/London, 1994) by Portishead, a Bristol band active in the 1990s whose members are often described as pioneers of the lo-fi musical movement or style referred to as "trip hop." Our focus will be on the ways in which these musicians explore predigital media signatures while simultaneously positioning themselves as avid practitioners of digital mediation.

Digital Silence and the Holy Grail of "High Fidelity"

Thanks to its historical impossibility, total silence is one of the most characteristic signatures of digital mediation. It looms large in the context of the cultural and historical quest for complete transparency in the technical mediation process—that is, the holy grail of "high fidelity" (hi-fi), or accuracy to the sound source.[1] The term itself was first coined in the late 1920s or early 1930s,[2] but a concern with the fidelity of sound arose even earlier, as we can see in a 1915 publication for Edison dealers:

Handling a Customer in the Store

Shopper: Do you claim to have something better than the Mineola?

Mr. Brown: Comparisons are always odious. The Mineola has no superior—in the class to which it belongs. The Edison Diamond Disc is a more expensive instrument and in quite another class.

Shopper: Is the Edison tone equal to the Mineola tone?

Mr. Brown: The Edison has no tone.

Shopper: No tone?

Mr. Brown: Exactly that. Mr. Edison has experimented for years to produce a sound re-creating instrument that has no tone—of its own. ... If a talking machine has a distinctive tone, then such tone must appear in every selection, whether band, orchestra, violin, soprano, tenor or what not. In other words, there is a distortion of the true tone of the original music.[3]

This claim of the complete erasure of the phonographic medium's self-presentation today appears preposterous; when listening to recordings from this early period, we are as likely to focus on the sheer level of noise and poor sound quality as we are on the music itself.

The noise of the phonograph, of course, derives from its very construction. A recording horn captures and concentrates the performed sounds, and a diaphragm (membrane) placed at the end of the horn vibrates in response to the sound waves (just like the ear's tympanic membrane). A stylus (pointed tool) that is connected to the diaphragm moves in line with the diaphragm's vibrations, cutting a groove similar to the vibration patterns of the actual sound waves into a wax cylinder or disc. The phonograph thus relied on the conversion of one motion into another via physical contact, and it was that contact (between stylus and cylinder) that produced the significant background noise. It was not only the noise, however, that limited the phonograph's sonic fidelity to its sources—sounds that were too loud could make the recording needle jump and consequently damage the wax, and sounds that were too quiet would be missed altogether. In addition to this limited dynamic range, the phonograph could accommodate only a limited frequency content. While the human ear is capable of perceiving sound frequencies between 20 Hz and 20 k1Hz, acoustic recording could only capture and reproduce sound frequencies between 168 and 2,000 Hz (Day 2000, 9–12). Given these limitations, recording engineers experimented with the positions of the instruments and singers in relation to the recording horn and even adjusted the musicians' styles to produce more audible dynamic shifts. Sometimes they modified the instrumentation itself: the felts of a piano's hammers could be filed down, for example, to make the sounds piercing enough to be captured by the recording device, and a tuba could be substituted for a double bass for the same reason.[4] Thanks to the phonograph's narrow frequency content and dynamic range, a performance that was originally rich and powerful could sound poor and thin on wax. Eventually, this combination of characteristic background noise and limited sonic range came to constitute the sonic signature of the phonograph as a recording medium.

This sonic signature of the phonograph became truly apparent in the context of the medium's successors, the first of which was electromechanical recording, which became the standard recording format after 1925. This recording medium replaced the recording horn of the phonograph with a condenser microphone that converted sounds into electric currents, and those electronically amplified currents, not a diaphragm's vibrations, drove the movements of the stylus, or cutter (Millard 2005, 141). Nevertheless, the physical contact between stylus and cylinder remained, and therefore so did the background noise. But the advantage of transforming sounds into electric currents lay in the process's enhanced capacity for amplification, which in turn broadened the recording medium's capabilities for capturing the dynamic range of the sounds. Electromechanical recording also captured a wider frequency spectrum than its predecessor, which meant that instrumentation required less tweaking as well (Day 2000, 16, 33). Very early on, electromechanical recording could capture frequencies between 100 and 5,000 Hz, and by 1934 the device was able to capture frequencies up to 8,000 Hz (ibid., 16–19). This enhanced spectrum made the recorded music sound fuller, brighter, and more realistic, and consequently the sound of the recordings became more and more true to the sound of the performance as experienced in the concert hall. These improved capabilities (coupled with the still persistent background noise) came to characterize the sonic signature of the electromechanical recording medium.

It was the background noise of the phonograph that ultimately compelled American engineer Oberlin Smith to search for a recording method that was *not* mechanical (that is, dependent on physical contact between stylus and cylinder). Smith identified the basic principle behind magnetic recording in 1878, but the magnetic tape recorder's entrance into music recording studios did not take place until 1947, and it only reached the mass market in the early 1950s. Despite Smith's original motivation, however, the reproductions generated via magnetic tape initially did not improve upon the sound quality of electromechanical recordings (though there were other advantages, such as the new editing abilities it offered). The analog recording medium still left unwanted noise on the recordings, such as tape hiss and crackle. (It was not until 1966 that the British Decca studios introduced a noise reduction system that added ten decibels to the music, so that the background noise was less noticeable [Day 2000, 21].) In the early 1940s, its frequency response was 50 Hz to 10 kHz, and the dynamic range was 60 dB (Engel 1999, 64); by the end of the same decade, the frequency range had been increased to between 30 Hz to 15 kHz (Gooch 1999, 86).

In addition to the fact that the *recording* media of these eras caused signal interference, the *playback* media added extra noise to the already degraded sounds as well. Edison's wax cylinder, which was the standard recording *and* playback medium until 1910, the 78 rpm shellac disc, which was the dominant playback format between 1910 and about 1950, and the vinyl record, which was, together with the compact cassette, the dominant playback format throughout the second half of the twentieth century, were all constructed around the same principle: the grooves engraved in the phonogram (a recording process we described above) are converted back into sound signals by means of a stylus that vibrates while tracking the grooves during a steady-speed rotation. Since all of these playback media rely on physical contact between the audio information (the engraved grooves) and the encoder of this information (the stylus), they undergo wear and tear each time the stylus travels through the grooves, which gradually degrades the sound quality of the disc or cylinder. Moreover, the physical contact between stylus and disc generates frequency hiss (the sound level of which depends on the medium), and the grooves on the disc or cylinder surface also attract dust and dirt, which can produce popping and ticking sounds. The tape of the compact cassette, the mass production of which began in 1964 (Morton 2006, 161), also produced unavoidable noise in terms of "tape hiss"—that is, high-frequency random noise.

While the mechanical, electromechanical, and magnetic recording and playback media all colorize the recorded sounds significantly in terms of various forms of noise, as well as their respective limitations on dynamic and frequency content, these sonic traces are eliminated by the digital recording and playback medium. This new technology achieved a dynamic range up to 98 dB, and a frequency range that included all of the frequencies audible to the human ear (20 Hz to 20 kHz).[5] Moreover, since the sounds are converted into numbers, the digital recording medium adds no noise whatsoever to the sounds.

The digital *playback* medium, the compact disc (CD), which was introduced in 1982 but achieved commercial success only in the early 1990s (Morton 2006, 172–173), also achieved total transparency or silence by relying on a laser optical decoder to read the binary codes from a disc surface made of polycarbonate plastic, in which the 0s are represented by pits and the 1s are smooth. The laser beam moves across the disc and the light is reflected back by the pits in different ways, which allows a light-sensitive photodiode to convert the codes into electric currents (Millard 2005, 350–351). Since there is no physical contact between the coded disc and the decoder, this technology avoids the deterioration over time that is caused by the playback medium (as we will discuss in the next chapter, the CD might, of course, nevertheless be damaged by other means), in contrast to previous playback media.

Thomas Greenway Stockham, who developed the first commercial digital-audio recording system, has, according to Greg Milner, stated that what he regarded as the most "surprising" feature of digital audio "is that, except for the use of a finite number of digits to represent each sample, the reconstructed audio can in theory be made to be *exactly* the same as the original" (quoted in Milner 2010, 222). While this claim might, of course, be disputed from an audiophile's perspective, in the sense that we can always achieve better sound quality, it represents, at a minimum, the way in which the average listener of our time hears digitally produced sounds. The way we hear digital sounds and silence, that is, must be understood within a historical context and therefore cannot be explained in terms of technological details alone. In the iterative history of sound quality, digital's transparency and silence marked the Holy Grail of "high fidelity."

Past Medium Signatures Revisited

While many manufacturers, music makers, and consumers celebrated the digital's purported achievement of "perfect" fidelity of reproduced sound-to-sound source, others took a more nuanced view. Andre Millard explains: "The almost clinical reproduction of the CD took some getting used to. Its range of frequency response matched that of the human ear, and it reproduced both highs and lows so exactly that it brought new meaning to many old recordings. It was uncommonly clean sound, so pure that it initially battered eardrums used to the comforting hiss of tape and the blurred, partly obscured highs of the vinyl disc" (Millard 2005, 353). Albin J. Zak III points out that digitally recorded sounds were often criticized for sounding "harsh," "brittle," or "cold," noting that, in a sense, the digital's cleanness could be regarded as distortion of sorts in itself (Zak 2001, 113). Such a sentiment informs the liner notes of the CD *The Rich Man's Eight Track Tape* (Homestead Records, 1987) by the alternative rock group Big Black (led by the now influential rock producer Steve Albini): "This compact disc, compiled to exploit those of you gullible enough to own the bastardly first-generation digital home music system, contains all-analog masters.

Compact discs are quite durable, this being their only advantage over real music media, you should take every opportunity to scratch them, fingerprint them and eat egg and bacon sandwiches off them. Don't worry about their longevity, as Philips will pronounce them obsolete when the next phase of the market-squeezing technology bonanza begins."[6] The remastering of vinyl records into the digital format of the compact disc further stirred up the debate about analog versus digital, and, according to Milner, some bands, such as the Rolling Stones, initially refused to remaster their music, claiming that "the grittiness of tape was integral to their sound" (Milner 2010, 209). Similarly, Neil Young once declared the LP format to be integral to his music and consequently dismissed its reissue on CD as a sort of forgery: "My album *Everybody Knows This Is Nowhere* is now available on CD, but it's not as good as the original, which came out in 1969. Listening to a CD is like looking through a screen window. ... It's an insult to the brain and the heart and feelings to have to listen to this and think it's music" (quoted in Gracyk 1996, 22).

The reason some musicians and producers in fact preferred the analog equipment was not that they experienced its sound quality as more trustworthy or transparent to the source, but the opposite—they favored it because of its characteristic signature, which they heard as an extra aesthetic dimension and "warmth" in the sound. Digital recording left the bare sounds completely exposed. Interestingly, it also occasioned a renewal of interest in how analog "clothed" those sounds, as an opportunity, this time, rather than a burden.

According to Jonathan Sterne, Edison's tone test managed to convince listeners of the phonograph's fidelity to the source despite its significant noise because they were able to separate the internal sounds from the external, privileging the music over the background noise of the recording and playback media (Sterne 2003, 262). Despite its sonic limitations, which must have been clearly audible even to its contemporary consumers, the phonograph's new means of capturing sound seems to have overshadowed its drawbacks. Faith Stone (the first wife of Compton Mackenzie) evocatively describes her introduction to the electrical amplified phonograph in 1928:[7] "[I] found this new noise quite unbearable, though it was the latest thing and first-rate of its kind. ... Amplification spells vexation ... but I soon got used to it, and even enjoyed it, which may or may not be a good thing" (quoted in Eisenberg 2005, 91). Listeners soon learned not to focus on this specific sonic imprint of the medium but on the music that the medium represented; they more or less acquired what Rick Altman labels "selective deafness" (Altman 1992, 30). While listeners were used to listening *through* the medium signatures of musical equipment and recording technology, digital technology's elimination of sonic colorization demanded that they start listening *to* these sounds. In "Making Old Machines Speak: Images of Technology in Recent Music," Auner points out: "When technology is replaced the limitations come to the fore; the veil of transparency is lifted and we are forced to start listening to the accent as all the repressed characteristics of the old emerge with shocking clarity" (Auner 2000). We have in previous chapters pointed out that opaque mediation over time might turn into transparency, but here is an example of the opposite: what was once taken to be almost transparent was suddenly noticed anew as quite opaque.

It was not only that the noise of predigital media became more "present" upon the arrival of digital silence, however—the meaning and function of the medium signatures of the predigital media changed as well. These things ultimately depend on how we listen to the sounds in the first place; Andrew Goodwin illustrates this point by observing that some musicians in the 1970s regarded synthesizers as "cold," mechanical, and artificial, while postdigital musicians just a decade or two later described the same synthesizers as "warm" and authentic, or expressive of a "human feel" (Goodwin 1990, 265). Just as the new is always heard in light of the old, the old will always be revisited in light of the new.

As part of the countercultural reaction during the 1990s to the promotion of digital technology's "victory" over low fidelity, several musicians made recordings during this time that featured the sound of predigital recording and playback media, and predigital instruments and other music equipment. Amplified vinyl noise, for example, can be heard on Tricky's "Hell Is Round the Corner" (*Maxinquaye*, Island, 1995), DJ Shadow's "What Does Your Soul Look Like" (*Entroducing*, Mo' Wax, 1996), Alanis Morissette's "Can't Not" (*Supposed Former Infatuation Junkie*, Maverick/Reprise, 1998), Massive Attack's "Teardrop" (*Mezzanine*, Virgin, 1998), Moby's "Rushing" (*Play*, V2 Records, 1999), and numerous tracks by Portishead, to whom we will return shortly.[8] Musicians and producers in the 1990s lo-fi movement produced the sounds of predigital technologies and techniques "naturally" by using these technologies, or they sampled the old and noisy sounds from existing recordings, or they could reconstruct them anew. In 1998, *Future Music* published a two-part guide called "Go Lo-Fi! Back to the Future" in its February and March issues that described methods and equipment that would contribute to "gritting up" the sounds. *Future Music* journalist Matt Thomas starts the first part of the guide as follows: "Sick and tired of spacious reverbs and 24-bit delays? Bored of subtle compression and glossy mixing? Then come with us on a journey into sound. Monographic crap sound" (Thomas 1998, 78). The ironic undertone of these introductory phrases is made obvious from the succeeding texts, in which Thomas makes clear that he prefers these "crappy" sounds, with phrases such as "that's the annoying thing about digital delays, they're too bloody good" (ibid.). The second part of the guide, written by Thomas and Dave Robinson, reviews some of the new signal-processing effects that can re-create the sound of old equipment: "For the first time in the history of recording there is equipment that intentionally makes your sound worse," they explain (Thomas and Robinson 1998, 80). In addition to popular "retro-grot" sound effects (their term) such as Lovetone and Mutronics, they mentions the DAW plug-in BIAS SFX Machine, which, among its over two hundred built-in presets, offers simulations of bad speakers (such as telephones or car stereos), vinyl crackle, and radio tuning drift. They also single out Steinberg's Grungelizer plug-in, which makes the music sound like an old record by turning different knobs to control record crackle according to parameters such as age and rpm (ibid., 80–84). Already in the late 1990s, then, there were available, in Stan Link's words, "some very high-tech means to achieve 'lo-fi' ends" (Link 2001, 35).

Vinyl noise and other sounds of predigital technology are now very common in contemporary popular music production, but as part of the music maker's sonic palette, not as a casualty of the available technology. Portishead, as mentioned, was among the 1990s bands that made use of this wide repertoire of sonic signatures from earlier times by sampling old sounds (and sometimes highlighting the medium signatures of those samples, such as their hum and crackle or limited frequency range), using predigital sound equipment, or exploiting the appropriate signal-processing effects to create similar sounds. What makes Portishead's music particularly interesting here is that, even as they plunder the analog past, they overtly embrace the digital present as well.

Portishead's Juxtaposition of Old and New

Portishead is Geoff Barrow (keys and programming), Adrian Utley (guitars and programming),[9] and Beth Gibbons (vocals); along with Massive Attack and Tricky, among others, they are often described as pioneers of the trip-hop movement that took place in the early 1990s.[10] Several of the artists since identified with this subgenre of electronica prefer the label "Bristol sound," since it was in this port city on the west coast of England that this particular musical style emerged. As trip hop came into its own, rave culture was sweeping across Britain and hip hop reigned in America. The "hop" of "trip hop" might acknowledge the fact that the music often retains hip-hop-inspired breakbeat rhythms, looped samples, and scratching. The "trip," on the other hand, might refer to the music's "trip" through diverse styles (techno, house, jazz, dub, reggae, and soul), or, more likely, to the "trip" associated with drug use, and especially marijuana—some claim that the mood, the sound, and even the rhythms of trip hop are explicitly designed to evoke the effects of this drug.[11]

Trip hop can sound rather narcotic, with its mellow strings and use of the synthesizer pad (a sustained chord or tone with very long attack and decay time), gentle brush beat on the drums, and drifting basslines behind a passionate female torch singer. On the other hand, it can be loud and intense, with scratching turntables, bass-driven hip-hop beats, and thumping basslines accompanying a grunting rap vocal. Given all of this variety, the most characteristic aspect of trip hop may in fact be the particular ways in which the music foregrounds its mediation, in terms of vinyl crackle and tape hiss, distorted sounds, bit-reduced, old samples, narrow-frequency sounds, old analog synthesizer use, and predigital sound-processing effects (see Brøvig-Andersen 2007). In other words, trip hop's style derives directly from the ways in which its artists revisit, experiment with, and juxtapose the characteristic medium signatures of diverse musical eras.

In interviews, the members of Portishead have voiced their obsession with the sounds of yesterday, and according to *Future Music* journalist Derek O'Sullivan, Adrian Utley "is renowned for his collection of classic guitar amps and effects boxes, incorporating everything from analog tape echoes and old Electro-Harmonix fuzz boxes to temperamental amps like the Ampeg Reverb Rocket and, his latest acquisition, a Leslie rotary speaker" (O'Sullivan 1998, 76). With the pronounced lo-fi attitude underpinning their stylistic mingling of hip hop, acid house, ambient music, cool jazz, and melodious pop, Portishead's sound is often associated with "telephone voices," vinyl crackle, tape hiss, old samples, and old instruments such as the theremin. Although the members of Portishead are obviously fascinated with the sound of old equipment, Utley acknowledges their debt to the digital medium: "None of us could say we hate digital because absolutely everything we do ends up in a sampler, which is digital" (quoted in Curwen 1999, 74). To create their commercially successful 1994 debut album, *Dummy*,[12] the band members apparently used a sampler connected to a computer (O'Sullivan 1998, 76; Miller 1995). This was before audio recording had become a commercially available feature of computer-based sequencer programs (discussed in chapter 1). Portishead, in fact, often recorded tracks first to analog tape and then sampled them from the tape in order to manipulate them (Curwen 1999, 75). They evidently used the computer-based sequencer program connected to the sampler to arrange and edit the sounds that would ultimately comprise *Dummy*.

"Strangers," which appears on this debut album, starts with a five-second sample from the introduction of "Elegant People" (*Black Market*, Columbia, 1976) by the jazz-fusion band Weather Report. The sample is easily recognizable despite the fact that it has been pitch-lowered and slowed down. Portishead has also added significant vinyl noise to the already existing medium distortion of the sample. After this short intro, a grainy drum loop starts up, together with a regularly accented and distorted beeping-alarm sound, additional vinyl noise, and an atmospheric, low-pitched drone. The heavily compressed and throbbing drum loop sounds as if it too has been pitch-lowered and slowed down. After twenty-eight seconds, this crackling, grungy atmosphere of vinyl noise and distorted sounds is suddenly replaced entirely by a clean, reverb-drenched acoustic guitar accompanying Beth Gibbons's melodious vocals by playing a minimalistic samba riff. In the background, we also hear some extramusical sounds at this point, as if there is someone in the room with the band, which associates the music with a typical live music session. Thanks to the acoustic guitar and the intense, expressive vocal performance, we might start to imagine a performance in the vein of the singer-songwriter tradition, but not wholeheartedly: Portishead has filtered out the high and low frequencies of the voice here, producing a sound that belongs to early recording and playback equipment, a transistor radio, or an old telephone (think of the familiar "telephone filter"). After a verse, there are two seconds of digital silence behind a solo synthesizer that supplies three pulsating and distinct tones (1:12–1:14), then the distorted drum-and-alarm groove takes over to accompany the vocal. Portishead removes the vocal filter at this point, so the play between clean and "dirty" sounds persists, but it has now been revised; either the vocal is clean while the instrumental sounds are gritty or the other way around. Soon, the drum-and-alarm groove is suddenly interrupted again, this time by what sounds like a sample of an old orchestra recording—its sonic characteristics stick out markedly from the rest of the music (2:00–2:12). The distorted drum-and-alarm groove takes over again but soon gives way to a one-second signal dropout that results in total digital silence (2:21–2:22). Then the groove resumes as an accompaniment to the vocal, drops out for a moment (leaving the vocal in complete digital silence with only a reverb tail and a few distant springlike sounds), and reappears again until the track abruptly stops.

The Weather Report sample that introduces "Strangers" is from 1976 but sounds as if it belongs to an earlier era, thanks to its limited frequency range and high level of background noise; it was likely manipulated by Weather Report and then again by Portishead to achieve this effect. Portishead's interest in sampling from old records is evident in several of their tracks: "Sour Times" contains a 1954 sample from Lalo Schifrin's "The Danube Incident"; "Biscuit" contains a 1959 sample from Johnnie Ray's "I'll Never Fall in Love Again"; and "Glory Box" contains a 1971 sample from Isaac Hayes's "Ike's Rap II" (all from *Dummy*, 1994). While the practice of sampling often represents a historically and culturally conditioned engagement with one's musical roots or an act of homage to earlier artists, Portishead also appears to sample out of their fascination with the characteristic *sound* of a particular era.[13]

As mentioned, the sequence between 2:00 and 2:10 in "Strangers" sounds like a sample from an old vinyl record of classical orchestra music. However, the source of this sample is not specified in the liner notes, so this orchestral sequence is probably not preexisting music at all but a new composition replete with a reconstructed medium signature of a long-ago era. In this case, then, the members of Portishead have sampled themselves, in a sense, in the guise of something else.[14] Likewise, "Western Eyes" (*Portishead*, 1997) purportedly contains a 1957 sample from the Sean Atkins Experience's "Hookers and Gin," but, according to online forums and our own Web search, this album does not exist, nor does the band; it seems Portishead recorded this music and made the sound of it simulate a 1950s recording (some have argued that "Sean Atkins" actually denotes Shaun Atkins from the Bristol band the Whores of Babylon, who, according to the same sources, performed the vocals of this musical sequence; see, e.g., Ben 2006). Portishead have stated that they often sample vinyl crackle from actual old records and apply it atop their own recordings, and that they use processing effects or lo-fi recording gear to make their performances sound gritty and old.[15] Elsewhere, for example, Portishead indicates that the string orchestra on the track "Humming" (*Portishead*, 1997) was recorded in the London-based AIR Studios (Associated Independent Recording Studios, founded by George Martin and John Burgess in 1965), then transmitted to compact cassette to degrade the sound signal, and finally sampled from there and reinserted into the music. Guitarist Adrian Utley adds: "Me and Geoff [Barrow], in the studio, will have an idea for the end kind of result and the whole basis of what we do is making it sound like old breaks, old records or whatever it is" (quoted in Curwen 1999, 75).

The drums in "Strangers" also sound as if they are sampled, not only because they are looped but also because the sound quality is clearly predigital. According to the liner notes, however, they are not sampled; session player Clive Deamer (who has collaborated with Portishead on several tracks) played the drums for the track. Utley clarifies the band's approach here: "We'll get Clive in to do a drum session for a day and make loads of different loops, and then it's down to Geoff and me coming up with ideas and Dave [McDonald] will be there to record things. We'll be looking for a couple of loops to make a song up. We're very anal about how we make those sounds—we'll get a loop and we'll work on whether you hear something on it ringing over from the previous bar (the one that wasn't sampled), like the tail of something played before, which adds a different sound to the loop" (quoted in Curwen 1999, 75). Instead of asking a drummer to accompany a particular track, they ask for a range of different beats, which they then sample and insert in different tracks, undermining any sense of their actual "liveness." In addition to revealing a heavy vinyl crackle, the sound quality of the drum loop in "Strangers" is just generally poor. This might be partly the result of using a 12-bit sampler (Adrian Utley explained in an interview from 1995 that he used an Akai 950, which he preferred because of its "grainy sound" [Miller 1995]), or partly the result of reducing the sample bit standard even further, to 8-bit, to degrade the sound quality.[16] As mentioned, it also sounds as if the drums are lowered in pitch and slowed down, both of which also suggest the analog medium: when a sample is given a slower tempo using an analog medium, its pitch also automatically changes; in the digital medium, on the other hand, these two parameters can be treated independently. The tempo reduction results in an extended duration for each attack and decay of the drum sounds, and the pitch lowering, of course, increases the activity in the lower frequencies and reinforces the dirty or gritty character of the drums.

Even as the predigital sounds imply the music's alignment with another era, the moments of complete digital silence between the three synthesizer tones in the section between 1:12 and 1:14, as well as in the section between 2:21 and 2:22, reveal where the music *actually* sits in time, placing us firmly in the present even as the music speaks to (and in the voice of) the past.[17] The first of these two sections is represented in the spectrogram depicted in figure 4.1.

The frequency-filter changes applied to the voice in "Strangers" also signify that the music shifts between medium eras, and, similarly, the crackling, gritty drum-and-alarm sequence seems to belong to a different era

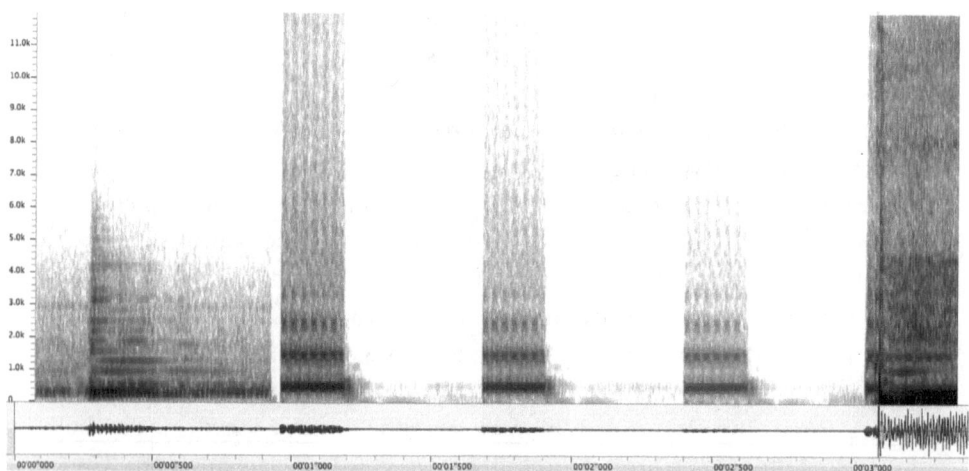

Figure 4.1
Spectrogram (0–20,000 Hz) showing digital silence between 1:12 and 1:14 in Portishead's "Strangers" (Amadeus Pro 1.5.4). Digital silence indicated by a complete lack of color (white areas) representing frequency response.

from that of the clean acoustic guitar. Most importantly, we are made to understand all of these juxtapositions as conscious aesthetic choices rather than unavoidable technical limitations, based on the music's complementary assertion of the digital as well as Portishead's back catalog. O'Sullivan sees this juxtaposition of signatures and styles as a fundamental characteristic of Portishead's music: "Grainy, shuffling drum grooves, pulsing compressed bass, moody drops and scratches and brooding aural peripherals, all sitting under a production hallmark that manages to instill classic 'oldness' with the clarity and polish that's become essential for contemporary, and—let's not deny it—radio friendly music" (O'Sullivan 1998, 76).

This mingling of digital silence and analog noise is also evident in some of the songs from their next album, *Portishead* (Go! Discs/London, 1997), most notably in "Undenied," "Over," and "Elysium." "Undenied" starts with a Rhodes riff, playing in an atmosphere of digital silence for about twelve seconds, at which point the silence is supplanted by amplified analog noise. The silence then reemerges for two seconds before the first verse; it later replaces the noise completely during the third verse as well as in the outro of the song. Like "Undenied," "Over" starts with an instrumental intro (this time, a guitar riff) atop digital silence, which supplies the background to all of the sounds until the second verse, when analog noise replaces it. Except for two short sequences of digital silence (2.13–2.15 and 2.43–2.54), the analog noise persists until the outro of the song, when the digital silence reemerges to frame the analog noise from either end of its progression. In "Elysium," there are several short moments of digital silence (00.33–00.36, 00.57–01.00, 1.22–1.24, 1.54–1.59, 2.09–2.12) that interrupt the otherwise noisy backdrop of the song. These small moments of digital silence anticipate the interlude from 2.48 to 4.12, which consists of guitar and piano playing in a completely digital climate. When Portishead mixes and contrasts new and old technology in a single track, we hear both media anew, an effect that would be lost if one of the media were to last from beginning to end.[18]

Portishead's music might evoke Fredric Jameson's description of the postmodern pastiche as an "imitation of dead styles" or a "cannibalization of all the styles of the past" (Jameson 1984, 65–66). Yet, by putting their vinyl noise within a digital frame, the band members position themselves not outside history but solidly within the present. This interest in sounds from the past and the various musical possibilities of lo-fi in tandem with or contrast to the digital, represented by the music of Portishead and other musicians of the 1990s, dovetails with the larger subcultural "retro movement," to which we now turn.

Retro Revivalism

The contemporary retro movement arose in the early 1970s and today extends from fashion to furniture, cars, social pastimes, and so on. In *Retro: The Culture of Revival* (2006), art historian Elizabeth Guffey describes the notion as follows: "'Retro' suggests a fundamental shift in the popular relationship with the past. ... Half-ironic, half-longing, 'retro' considers the recent past with an unsentimental nostalgia. It is unconcerned with the sanctity of tradition or reinforcing social values; indeed, it often insinuates a form of subversion while sidestepping historical accuracy" (Guffey 2006, 10–11). Building her theories on, among others, the postmodernist Jean Baudrillard (2004) and historian Raphael Samuel (1994), Guffey defines "retro" as a more distinct sensibility than these scholarly predecessors. Among other things, she finds that, unlike earlier forms of revivalism, retro focuses on the relatively *recent* past, rupturing historical continuity by positing a great divide between yesterday and today that in turn makes yesterday suddenly seem distant and old. Moreover, she observes that in revisiting "the outmoded and passé, retro claims a Camp sense of rediscovery" (Guffey 2006, 161). Instead of recollecting proud examples of the past, retro finds pleasure in highlighting its "weird" or "ugly" (or at least unexpected) features (ibid.).

This characterization of retro fits the ways in which Portishead and several of their contemporaries highlighted analog noise while framing it with digital silence. If, in the 1990s, analog noise was no longer considered an unavoidable feature of the present, it was also not yet considered to be something that belonged to a distant past, given that the CD format only took precedence over the LP and cassette formats in the early 1990s. In fact, listeners were at the time more familiar with analog noise than with digital silence. Nevertheless, the ways in which Portishead and their contemporaries *reframed* that noise by contrasting it with the silence of the digital medium, and then made the contrast even more profound by exaggerating the noise, made these very familiar sounds belonging to the very recent past suddenly seem completely "vintage." If the realization of the retro implies a fascination with the past's "ugliness," or at least highlights unexpected features of the past, these musicians' embrace of what used to be regarded as a technological *limitation* echoes the sentiment. Sounds that once were lamented were now valued and even exaggerated.

Of course, the exaggeration of analog noise distorts the reality of an analog recording, the noise of which is much more subtle. As Samuel argues in his description of "retrochic," retro actually involves a historicist fantasy that subverts historical accuracy while acknowledging its ironic stance (Samuel 1994, 95). While approaching the past with admiration, that is, the retro approach also incorporates an element of disloyalty and detachment. This ironic, often humorous interpretation of history and arbitrary recollection of the past comprise a uniquely unsentimental form of nostalgia. Contrary to those earlier forms of revivalism that are characterized by a nostalgic yearning for the past, retro fuses the old with the new, revealing its underlying commitment to the present. Retro's fascination with the past, in other words, does not imply a rejection of the present. While Portishead's affection for predigital sound might appear nostalgic, the band also obviously values the digital medium, because, as we've emphasized in our analyses of their music, they highlight its characteristics right alongside their (also digital!) play with the analog.

The dual foregrounding of predigital media signatures in a digital frame invites us to look at today's technology as a medium with its own voice. This last point returns us to Louis Marin, who concluded the same sort of thing in relation to the dark background of a painting, which he described as a nonrepresentational self-representation: "What does this dark background represent? Nothing. 'Nothing' names something which cannot be named. But if this background does not represent anything, it presents itself as nothing: it presents itself as not representing something. And it is this pure self-presentation that allows the whole painting to represent the three objects with such force" (Marin 1991, 66). Just as this "nothing" is in fact a form of self-representation, so is the completely silenced background of the digital format. Digital silence need not be understood only as "nothing"—as the erasure of medium materiality—but as a signature of its own, and one that is in fact worth "listening" to.

Auner notices something similar about Portishead's "Undenied": "The opposition of the sound of a very scratchy record and digital silence become an integral part of the composition" (Auner 2000). Moreover, he points out that while the digital is often described as "cold" and contrasted to the "warmness" and "emotionality" of the analog, "Undenied" reverses these stereotypes. By contrasting the here dominant sound of old technology with just a few bars of utter digital silence, it is the latter that delivers an extra emotional punch: "Here when the scratchy noises and cymbal hiss drop out we are confronted with a desperate emptiness. Through the lyrics the vinyl noise becomes the embodiment of the obsession; the thought of absence results in the moment of absolute emptiness represented by the digital silence, now made horrible and empty" (ibid.). The digital void of sound and noise, then, might function as a metaphor for emotional emptiness. During the moments of signal dropout in "Strangers," the background of "nothing"—the void of sound and noise—is suddenly foregrounded and presented as a musical feature that provides aesthetic and emotional pleasure in the same way as older technologies. Portishead, then, brings new meaning to both the out-of-date and the up-to-date.

This juxtaposition of sounds from different eras challenges both positivistic views of technology and progress *and* suspicions about innovation; this music does not represent one signature as any better or worse than the other but explores the aesthetic potential, as well as the boundaries, of both. The present is thus neither rejected nor celebrated as the culmination of technological progress. As Guffey explains: "Retro allows us ... to see the unshakable faith in Modernity as limited in historic scope. Subtly, retro reworking of the Modernist past ... helps to put the Modern in past tense" (Guffey 2006, 25–26). This sweeping cultural means of revisiting the past seems to express an ambiguity about the contemporary, in which neither the new nor the old is valued in its pure or unmitigated form. Instead, the new is pushed forward through the reinvention of the past from the position of the present. In this way, as suggested by the title of the "go lo-fi" guide (Future Music), it represents a path "Back to the Future."

The 1990s reappropriation of predigital technologies and techniques represented a form of spatiotemporal experimentation, in the sense that the characteristic material signature of earlier media was split from its source and used as a compositional brick within a later technological context to produce an aesthetic effect. Brian Eno points out that the characteristic "limitations" (his word) of a medium communicate "something about the context of the work, where it sits in time" (Eno 1999).[19] But these "limitations," as we have seen, can be constructed as well as inherent, telling us as much about the present as the past.

The act of revisiting past signatures or using outmoded technologies is, of course, not new; popular music has long been fascinated with lo-fi sounds and sounds from the past. A lo-fi aesthetic had already characterized the garage rock of the 1960s, the punk rock of the late 1970s, and the hip hop of the 1980s, but digital silence added an urgency to the movement that crossed genres and styles. The various medium "limitations" of the immediate past became, once again, revalidated in the age of their potential absence. As Link puts it: "Given the current state of technology, the significance of such noise increases in direct proportion to its avertability" (Link 2001, 35). Digital silence allowed for a contrast with the noise inherent in previous recording and playback media that had never before been possible, and thus suddenly made these sounds from the very recent past seem "old." And just as the past is reframed by this silence, the silence is reframed by the exaggerated analog noise; this contrast, in turn, makes us encounter both the past signatures and the digital silence anew.

John Cage's description of noise rings true with regard to the signatures of predigital recording media and music equipment: "When we ignore it, it disturbs us. When we listen to it, we find it fascinating" (Cage 1967, 3). Of course, this also applies to the digital signature of silence. The members of Portishead expose not only old technology but also the medium of their own time, and in this way they invest the latter with *new* meaning. In the end, they question the assumption that the sonic signatures of previous media possess inherent aesthetic and emotional qualities that the digital medium does not, and they lay bare our self-interest in deciding this one way or the other. That is, the meaning and function of a medium's sonic signature are discursively dependent; we hear it differently when we use it differently (when we can take it or leave it). The same sounds can be listened *through*, or listened *to*—they can be desired as well as regretted. What they say, in short, is what we make of them.

5 Cut-Ups and Glitches: The Freeze and Flow of Los Sampler's and Squarepusher

If the cut-and-paste tool was a woman, I would marry her.
—Kid Simius[1]

The ability to cut and paste recorded material with scissors accompanied the development of magnetic tape. While some musicians and engineers stuck with a more conservative approach, rejecting the medium's newfound ability to splice tape, others welcomed the possibility but used it in a discreet or entirely hidden fashion, to eliminate unwanted sounds or move a sequence from one take to another to make things sound better. Still others, however, such as participants in the early-1950s electroacoustic music scene, as well as the literary experimentalist William S. Burroughs,[2] took a more experimental approach to cutting and pasting, fragmenting spatiotemporal structure until the recording/production medium itself became opaque rather than transparent to the listener. These early cut-and-paste techniques were soon adapted to the field of popular music and culminated in digital sequencer programs.

Although few of the sonic effects caused by the cut-and-paste tool are new with the digital medium, they are nevertheless much more common there, thanks to digital's malleable nature and nondestructive editable environments. For example, it took American composer John Cage a full year to produce the cut-up tape montage *Williams Mix*—a juxtaposition of hundreds of spliced audio tapes—whereas it certainly takes a lot less time to produce a digital cut-and-paste montage.[3] For example, the two tracks that we will analyze in this chapter would simply overwhelm any sort of physical cut-and-paste operation in the analog medium. The fact that "undo" operations were not even an option with predigital recording equipment made experimentation with these sorts of processes even more time-consuming.

Digital sequencer programs also allow for more precision in cut-up operations, as one can zoom in on the sounds and surgically manipulate them at a microlevel. Whether the musical goal is to audibly reassemble tracks from different recordings, to discreetly juxtapose different takes of a single performance, to make "loops," to cut out unwanted noise and sounds, or to make sonic and rhythmic effects that expose the operation itself, the cut-and-paste tool has become fundamental to the production of contemporary popular music.

The vastly more efficient virtual cut-and-paste tool eliminated the extremely time-consuming process of physically splicing tapes. Moreover, musicians in the popular music scene of the second half of the 1990s also started to embrace the sounds of skips, stuttering, and signal dropouts as musical material in their own right. Such sounds both work musically *and* evoke malfunctioning technology, such as a CD player that has problems reading the information on a scratched disc—an unfortunately familiar sound to most of us—or a computer program that halts or freezes during playback of an audio file.[4] Glitch[5] sounds can be manufactured in different ways. Some artists have produced sonic skips and stutters by generating actual technological glitches. For instance, the Japanese artist Yasunao Tone (who was active in the Fluxus movement in the 1960s) and the German electronica trio Oval stuck bits of pin-punctured Scotch tape on the data side of the CD or doodled on the back of the CD with a pen (and recorded the result), in order to generate skipping and stuttering sounds.[6] Skipping and stuttering sounds are, however, much easier to achieve using the cut-and-paste tool than they are to generate on their own with a physical CD or other medium. As Caleb Kelly points out in *Cracked Media: The Sound of Malfunction* (2009): "It is seemingly far too easy to cause a CD to skip and stutter; it appears at times that the slightest scratch causes the CD to skip just at the best part of one's favorite track. The irony is that the deliberate forcing of a CD to skip is a very delicate operation" (Kelly 2009, 217). While scratches or marks on a vinyl disc will make the stylus of the turntable literally jump or slide across the physical grooves of the disc, a CD skip results only when the laser of the CD player reads the binary codes of the CD incorrectly. Any minor loss of data inscribed in the CD, however, is automatically overridden by the player's error correction system, and any major loss will mute the sound or stop the CD from playing altogether. It is the betwixt and between here that results in a sonic jump, skip, or hang (ibid.,

77). Contemporary musicians seem to be fascinated with the rhythmic and sonic complexity of glitches, but instead of privileging the chance indeterminacy of natural glitches (as Tone does, for example), they carefully manipulate these effects using the cut-and-paste tool or various "glitch" plug-ins that simplify the process of cutting and pasting even further.[7]

This chapter features an analysis of "La Vida es Illena de Cables" by Los Sampler's (Uwe Schmidt), as well as an analysis of "My Red Hot Car" (*My Red Hot Car*, Warp, 2001) by Squarepusher (Tom Jenkinson), in both of which the experimentation with and foregrounding of cut-ups and glitch sounds play a significant part. Each of these songs will be analyzed in terms of their exploitation of the cut-and-paste tool, in the interests of demonstrating that the foregrounding of this particular instance of mediating technology is a big part of the music's overall design. Through these analyses, we will furthermore discuss how the ability of the recording/production medium to transparently represent a coherent performance might be both confirmed and denied, and how the cut-up sounds involved in these tracks might be understood to be both unmusical and musical at the same time.

Music within the Music

There exists very little information about Los Sampler's and the group's one and only album, *Descargas* (Rather Interesting, 2000). According to the recording's liner notes, Los Sampler's consists of seven men with names of Latin American origin, and the band is produced by "Atom™"—one of the many monikers of the German music producer, or all-in-one-musician, Uwe Schmidt (famous for his Señor Coconut project). The fact that Schmidt is recognized for his musical humor as much as his musical talent might make us question whether Los Sampler's really is a Latin American combo or just one of Schmidt's many solo projects featuring guest performers, an assumption supported by the fact that the album is released on Schmidt's own record label. Elsewhere, Schmidt has stated that Los Sampler's is a "Latino-Fake combo," consisting of five Latinos (not seven, as the cover suggests!) who "came together to make music which they had in their heads but which none of them could produce alone, because they didn't have the equipment" (Señor Coconut 2015). According to the recording's liner notes, all of the songs are programmed but derived from jam sessions that are sampled and that supply the "spirit" behind the album. On many of his musical projects, Schmidt has first recorded jam sessions by musical ensembles and then reworked the music on his computer by chopping it up, reassembling it, glitching it, and inserting into it additional musical sounds and effects; it is safe to assume that this is the way *Descargas* was produced as well.

The aim of Schmidt's record label Rather Interesting is, according to its website, to expand the realm of electronic music by featuring the unexpected, and *Descargas* certainly does so (see Señor Coconut 2015). Like his music project Señor Coconut,[8] *Descargas* is an instance of what Schmidt himself has called "electrolatin" music—a combination of electronica and Latin music. The album cover art evokes the 1970s and features a photograph of the seven (fake?) members of Los Sampler's, all of whom appear to be Latin American and wear beige suits with red shirts and beige bow ties. From this image, then, we might well expect music that is more traditional than experimental, and we would be right, to an extent. The album is based on traditional Latin (especially Cuban) music forms, such as the "son," the "descarga," the "cha-cha-cha," and the "mambo," and the performances feature typically Cuban instruments, such as the Cuban *tres* (a guitar with three courses or groups of two strings, tuned in a major chord), the Spanish guitar (that is, the classical nylon-string guitar), the double bass, and percussion instruments such as the shaker, the claves (a pair of short thick clubs), and the bongo (an attached pair of open-ended drums). The clave rhythm—a five-stroke rhythmic pattern—is central to this music, as are rhythmic and melodic repetitions in general; male singers deliver the Spanish lyrics of the album polyphonically. *Descargas* does not attempt to represent the current Cuban music scene but rather gives a nostalgic impression of an exotic and authentic past, when music was relatively unmediated by technology and delivered by copresent musicians singing live and playing traditional instruments. However, *Descargas* is full of sonic glitches that operate in tandem with these traditional-sounding Cuban music performances. Interestingly, while we might expect that the Cuban music has been sampled and then warped, the Spanish lyrics reveals that the music was, in fact, created in order to be warped. For example, according to the translation in the liner notes of *Descargas*, the Spanish lyrics on "El Nuevo New Looks (Son Chueco Cerebral)" read: "This is the 'son,' which is super-warped, out of tune, without meaning and without message, just mental message." Similarly, the lyrics on "El Rey de las Galletas (Soy Yo) (Rumbo Galleton)" read: "I invented the 'galletas,' listen Señor, how they are crunching. I am the king of the 'galletas,' that's me." *Galletas* refers here to "those very short digital clicks you can hear all over the album." In what follows, we will look more closely at the fourth track on this album, "La Vida es Llena de Cables (Son Disco Duro)."[9]

The track starts with a Spanish guitar playing a traditional riff in the style of the Son Cubano (many Cuban music styles, such as Salsa Cubano, originate in Son Cubano). This riff is repeated two times and gives no hint that there is anything unusual about this seemingly traditional Latin American musical performance. But when the guitar riff is repeated a third time, it is warped by skipping and stuttering sounds. To illustrate our auditory analysis of how the cut-and-paste tool impacts this guitar track, we have marked the various cut-ups in the DAW program Logic Pro 9, as displayed in figures 5.2 through 5.4. By comparing waveforms of the unwarped guitar riff (as it is performed the first and second time, and displayed in figure 5.1) and the warped guitar riffs and reading them while listening to the audible information, we can tell where the track was chopped up, which sequences were copied, and how they were pasted back together.

The waveform of the riff as performed the second time is identical to that of the first riff (as displayed in figure 5.1) and is probably just a copy-and-paste repetition of it. The third time through, however, we hear a glitch in the guitar riff: part of the first tone is cut up and copied two times (the

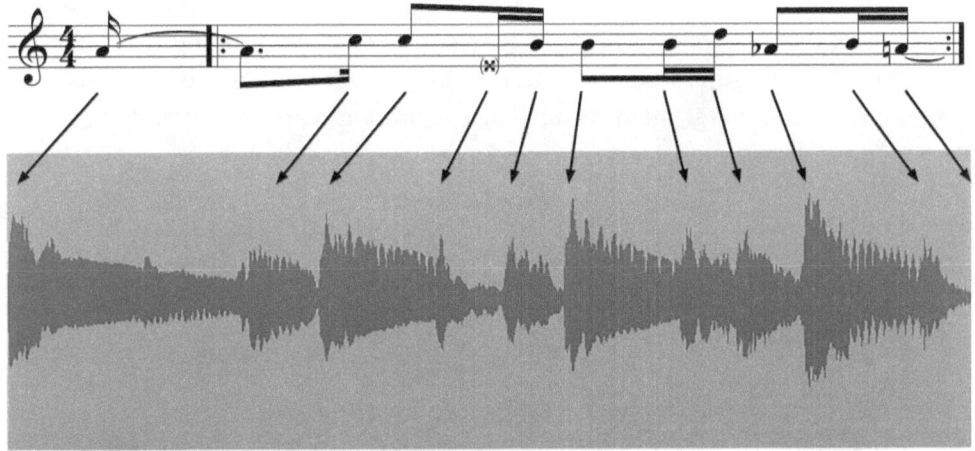

Figure 5.1
Transcription and waveform of the Spanish guitar riff as performed the first and second time in the introduction of "La Vida es Llena de Cables" by Los Sampler's.

copied sequences are marked with lighter gray in figure 5.2) to create a stuttering effect. The third tone is also copied, this time just once, and then followed by a very short, almost unnoticeable, signal dropout—that is, a sequence in which the sound signal disappears altogether. A very short fragment of the fifth tone is finally cut up and then repeated consecutively, producing several identical cells. These cells are repeated at such a short interval that, instead of a stuttering effect, they perform a percussive function that is comparable to programmed and quantized drum rolls.

The fourth time the riff is repeated, the first and second tones return in their unaltered states while the third tone is repeated and followed by a short dropout, as before. This fourth repetition contains several more signal dropouts as well (as displayed in figure 5.3). In addition to the almost unnoticeable dropout between the third tone and the ghost note, there is a short dropout between the ghost note and the fourth tone. The fourth tone is furthermore chopped in two, and the parts are separated by a short signal dropout, as if the sound has been disrupted by a passing technological glitch. After the fourth tone and before the fifth, a rather long dropout is inserted. Digital dropouts, of course, differ from traditional musical rests in that they are generally complete digital silence, without any form of atmospheric noise or "dead air."[10] Moreover, whereas traditional rests allow sounds to die out first, dropouts silence their sounds very abruptly ("too soon," in effect). This contributes to the impression that parts of the sound signal are somehow missing altogether—the result of, or at least a simulation of the result of, a technological malfunction. This fourth repetition of the riff ends with a stuttering effect on the seventh tone (part of which is repeated two times); an unmanipulated eighth tone completes the riff.

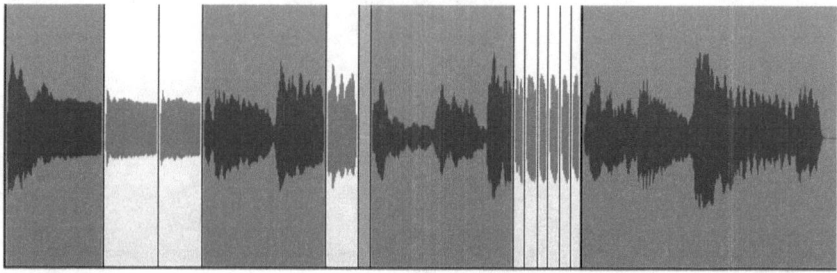

Figure 5.2
Waveform of the third repetition of the Spanish guitar riff of "La Vida es Llena de Cables" by Los Sampler's (Logic Pro X).

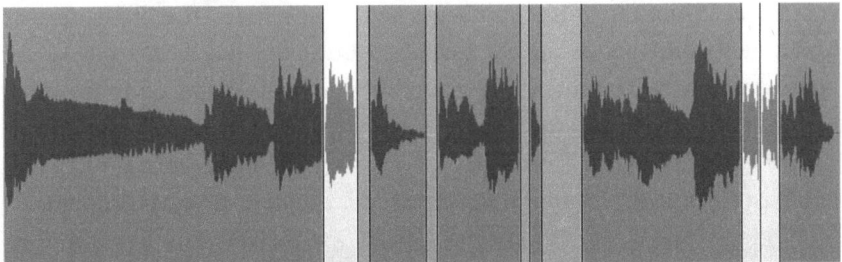

Figure 5.3
Waveform of the fourth repetition of the Spanish guitar riff of "La Vida es Llena de Cables" by Los Sampler's (Logic Pro X).

The fifth repetition of the guitar riff simply repeats the third repetition, but the sixth repetition has yet another glitch: a fragment of the first tone is replicated numerous times to form a series of identical cells, each of very short duration, so that the originally sustained tone is replaced with a "drum roll" of cut-ups (see figure 5.4). Like the third repetition of the riff, a part of the third tone is copied once and followed by a short dropout, and the fifth tone is cut up and copied several times, to produce a percussive effect. The seventh tone, however, differs from the third version of the riff, in that it is shortened and interrupted by the eighth tone (marked with lighter gray in figure 5.4), which again is prolonged through its manipulation into a quite discrete percussive effect.

The seventh repetition is identical to the fourth, and the eighth is identical to the sixth. After these eighth introductory bars, consisting of various versions of the same guitar riff, the Spanish guitar is joined by a double bass, Afro-Cuban percussion instruments, and the polyphonic male vocal

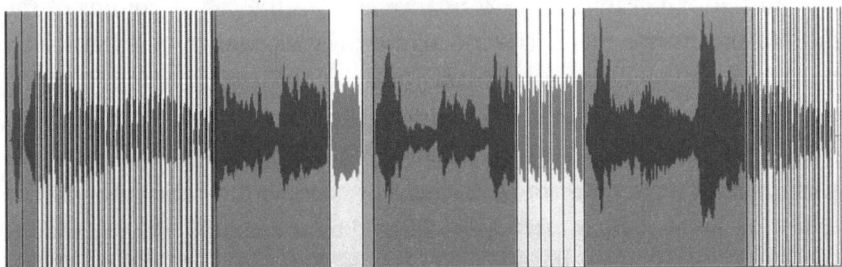

Figure 5.4
Waveform of the sixth repetition of the Spanish guitar riff of "La Vida es Llena de Cables" by Los Sampler's (Logic Pro X).

ensemble. While the guitar continues to sound glitchy, the other musical instruments and vocals appear unaltered until the music has played for approximately two minutes. After this, the "glitch virus" spreads to the other instruments as well, and the damage starts to escalate. From its beginning in the style of *Buena Vista Social Club*,[11] the song gradually devolves into a cut-and-paste montage of sound fragments and sequences of signal dropouts.

While we are, in our current mediatized culture, very familiar with fragmented music that exposes its technological mediation, fragmented music is a relatively new phenomenon on a historical-cultural scale, as is music recording and exact (mechanical) reproduction itself. The skips and stutters of "La Vida es Llena de Cables," then, are powerfully incongruent with our deep roots in understanding music as a spatiotemporally coherent singular performance (before recording was invented a century ago, music was always live and therefore only comprehended as such). Yet they do not *completely* dismiss this understanding, because the song is not based purely on technological glitches. In "La Vida es Llena de Cables," the audible cut-ups are staged as passing effects, between which we are meant to sense a coherent musical performance. The additional fact that "La Vida es Llena de Cables" suggests associations with the traditional Cuban music scene strengthens the presence of a familiar coherence underpinning the glitch experimentation; recordings in the traditional Son Cubano style are usually the result of, and in effect often smoothly conceptualized as the documentation of, a live performance in which the music is performed in a single spatiotemporal setting by a copresent ensemble and recorded in a single take. These associations are, for example, triggered by the fact that neither the guitar, the double bass, nor the percussion instruments sound "programmed." Thus, "La Vida es Llena de Cables" strengthens our expectations around traditional musical performance even as the cut-up sounds of the production disrupt them. The incongruity we experience here between a traditional Cuban music performance and the exposure of malfunctioning technology is even mirrored in the title of the album: *Descargas* might refer either to a jam session—which usually means a spatiotemporal coherent performance—or to the act of discharging, rejecting, or denying—an act that is in this particular instance performed by the malfunctioning sounds upon the spatiotemporally coherent performance.

Music such as "La Vida es Llena de Cables" disturbs the recording/production medium's realistic representation of a traditional performance at the same time as it underlines it; we would not experience these sounds as cut-ups, medium glitches, or signal dropouts if we were not aware that there was something to be glitched or missed. The cut-up techniques thus make the listener aware of the recording/production medium's double function, to mediate and to *be* that which is mediated—it presents itself while it mediates or represents something else. The conventions of the recording/production medium as a transparent mediator, and of music as a coherent performance, are at once "inscribed and challenged, used and abused," to borrow Linda Hutcheon's description of some postmodern art (which in fact can have many similarities to contemporary popular music in which the mediation is exposed) (Hutcheon 1989, 136). We are left to comprehend the song as consisting of two layers of music: the traditional and the manipulated. We might even end up with the impression of a recording that presents music within the music, as if the traditional performance constitutes the layer of music that is interrupted by the other musical layer of cut-ups and glitches. Using the theories of Albert S. Bregman, we could describe our experience of this music as consisting of different auditory *streams*. In *Auditory Scene Analysis*, Bregman (2001, 1–45) discusses how our brain structures complex auditory information into separate streams—for example, how it enables us to locate and identify a voice in a crowd or an instrument in a musical ensemble. He explains this process of grouping either as happening naturally and automatically (he calls this "primitive integration") or as happening on the basis of earlier experiences and learned restrictions (he calls this "schema-based integration"). The latter applies to the two layers we might perceive in "La Vida es Llena de Cables": because of our deep roots in conceptualizing music as a spatiotemporally coherent performance, we are likely to hear the music as consisting of a performance (one layer) and of sounds that are interrupting this performance (the second layer). Cut-up sounds, then, challenge our conceptions of what music is or could be in terms of disrupting the musical recording/production medium's realistic and transparent narrative or traditional representation of a spatiotemporally coherent performance.

The Double Meaning of Aestheticized Malfunctions

"La Vida es Llena de Cables" by Los Sampler's is part of the "glitch" movement of the late 1990s and early 2000s. "Glitch" comprises a particular music style within electronica characterized by its practitioners' exploration of the sounds of malfunctioning digital technology, often through their extensive use of the cut-and-paste tool.[12] Although musical glitches saw a revival through the subgenre of electronic music that came to be labeled "glitch," the aesthetic qualities of technological glitches, and their ability to undermine or reinforce existing aesthetic preferences, have been recognized for decades. For instance, the sound of malfunctioning technology played an important part in the avant-garde art movement known as Fluxus, the members of which were greatly inspired by the sonic performances of John Cage. Milan Knížák, a Czechoslovakian artist closely associated with Fluxus, is famous for his music project titled *Destroyed Music* (1963–79), in which he glued, among other things, four cut-up quarters of different vinyl discs together and played the result like a normal record (Kelly 2009, 144). In company with Swiss American artist Christian Marclay, he also anticipated both Tone and Oval's operations upon the CD by sticking various things to the surface of the vinyl record in order to produce abrupt transitions between sound sequences. In addition to *preparing* malfunctioning vinyl records, Marclay is known for his 1985 release *Record Without a Cover*, which was sold without a protective package or sleeve with the intention that the sonic result of the inevitable wear-and-tear damages (scratches, dust, and fingerprints) would be regarded as part of the work (ibid., 150–184). However, while glitch sounds once were associated with the avant-garde, the sound of digital glitches can now be found in more mainstream contemporary popular music as well.

The success of what *Wired* journalist Erik Davis calls "electronic pop stars" (Davis 2002), such as the Warp[13] artists Aphex Twin (Richard D. James), Autechre (Sean Booth and Rob Brown), and Squarepusher (Tom Jenkinson) (who all have a glitch-inspired cut-and-paste approach to their music), has contributed much to the popular appeal of digital musical glitches. In what follows, we will analyze the music of one of these three glitch-inspired "electronic pop stars," namely Squarepusher, in order to illustrate an even more extreme use of the cut-and-paste tool than is presented in "La Vida es Llena de Cables."

In 2001, Squarepusher released an EP consisting of four tracks as well as a bonus track. The first two tracks of the EP—"My Red Hot Car (Girl)" and "My Red Hot Car"—are quite similar, as the titles imply. In fact, the second track—subsequently placed on the Squarepusher album *Go Plastic* (Warp,

2001)—sounds as if it is simply a glitched version of the first track. This impression is confirmed when we examine the two tracks and compare their waveforms, as we will see in the following analysis.

The introductions to both versions of "My Red Hot Car" consist of drums (which sound programmed), a bass guitar, and a synthesizer. While the synthesizer in "My Red Hot Car (Girl)" plays a sustained chord lasting throughout the first 4/4 bar, this chord in "My Red Hot Car" is shortened to a quarter note and then repeated with a dropping volume level (as in an analog delay). When this chord appears a second time, in bar 5, it is shortened to the duration of a sixteenth note, followed by a signal dropout before it continues. (In the original track, the sustained chord is performed the same way as in bar 1 each time it returns.) At the end of bar 8 in the manipulated version, two short cut-up pieces from the synthesizer chord are inserted to anticipate bar 9, during which the chord is split up again and separated by dropouts; the same thing happens again (although in a different form) when the chord returns in bar 13.

After sixteen bars, the vocals appear and deliver the raunchy lyrics of the song: "You scream out for more / Let me tell you girl that's for sure / I'm gonna give you all I've got / I'm gonna fuck you with my red hot cock" (though the title would have it as "car," it does not sound like "car" on the recording). Even in the original version of the song—"My Red Hot Car (Girl)"—the voice sounds manipulated, in terms of being heavily autotuned as well as (much more subtly) filtered and distorted. Moreover, each phrase ends a bit abruptly, as if the vocals have not been allowed to resonate properly first. These manipulations are, however, taken to a new extreme in the second (*more* warped) version of the track. The vocals are all chopped up, and the sound pieces are often repeated as a stutter and/or separated by signal dropouts, which in turn overlay a more staccato rhythm upon the performance (see figure 5.5, which displays the waveforms of the original and the warped vocal performances of the first sentence). The cutting between or within words obscures them (and thus hides the text's meaning). They sound as follows (the strikethrough indicates a muted part of the word): "You-u-u-u scream ou-ou-out for mooooooore / Let me tell you girl-irl-irl that's foooooooor sure / I'm gon-na gi-i-i-i-i-i-i-i-i-i-i-i-ive you a-a-a-all I've-I've-I've-I've gooooooot / I'm gon-na fuck you with my re-re-re-re-re-red hot coooooooock."

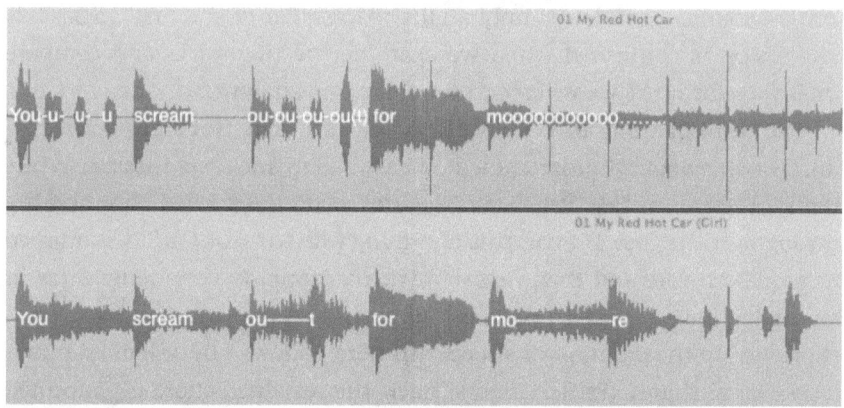

Figure 5.5
Waveform of the first sentence of Squarepusher's "My Red Hot Car" as it appears in (a) the warped version of the song, and (b) the original version of the song (Logic Pro X).

In the warped version of the first sentence, the "u" in "you" is cut up and pasted consecutively, in which each repetition is separated by a short dropout, creating a stuttering effect: "you-u-u-u." A dropout that consists of complete digital silence interrupts "Scream" halfway along, so that the pronunciation of the letter "m" is muted. "Out," like "you," is chopped up, copied, and pasted consecutively, producing a staccato stutter. On the word "for," the rising amplitudes before the peak of the attack are cut off, resulting in a very abrupt start. The first half of the last word of this first sentence, "mo(re)," is prolonged by copying the "o" but with smooth transitions, then drenching it in reverb while the volume gradually drops. The vocal's delivery of the second sentence follows in a similar manner (see figure 5.6, which displays the waveforms of how the second sentence is delivered by the vocal in the two different versions of the song).

While there are gliding transitions among the words "let me tell you girl" in the original version, the warped version chops up the transitions into a staccato rhythm, separated by short dropouts, during which each sound starts and ends abruptly. In addition, "girl" ends with a stutter: "girl-irl-irl." A distinct dropout is inserted between "that's" and "fo-o-o-or" as well, making "that's" end halfway along while "for" skips its first consonant. The "o" in "for" is repeated three times, while nothing is done to "sure" except that it is prolonged, drenched in reverb, and dropped in volume, as is "more" in the previous sentence.

Cut-Ups and Glitches

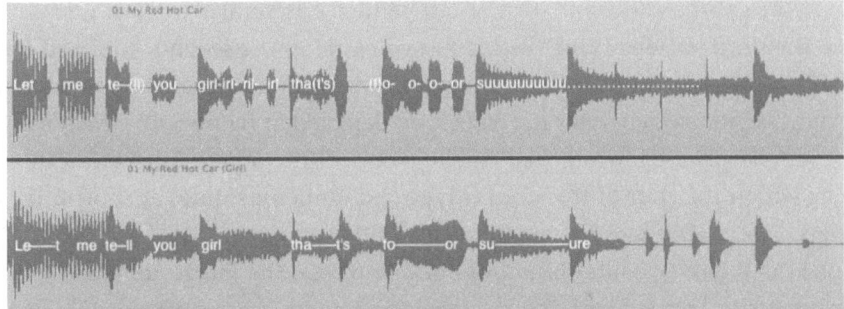

Figure 5.6
Waveform of the second sentence of Squarepusher's "My Red Hot Car" as it appears in (a) the warped version of the song, and (b) the original version of the song (Logic Pro X).

As with "let me tell you girl" in the second sentence, "I'm gonna" in the third sentence is chopped up into a staccato rhythm with short dropouts between the bits, giving each sound a sharp beginning and a sudden ending (the two different versions of the third sentence are illustrated in figure 5.7). The "i" in "give" is then repeated numerous times at such short intervals that it produces a percussive "drumroll" effect rather than a straightforward stutter. The "-ve" of "give" is replaced by a dropout and followed by a stuttering "you a-a-a-all I've-I've-I've-I've"; the sentence then ends like the last two, with "got" hanging in a reverberant atmosphere before the air absorbs it.

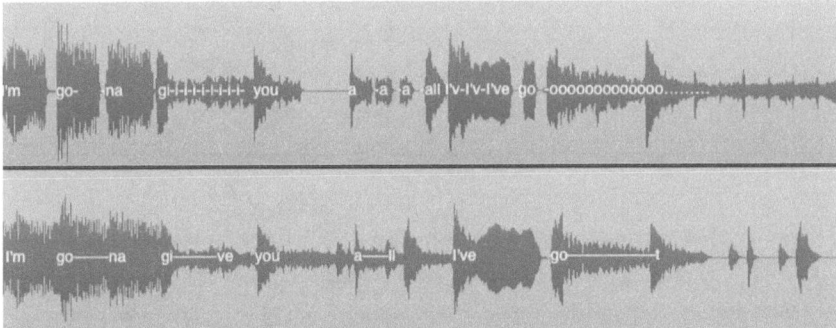

Figure 5.7
Waveform of the third sentence of Squarepusher's "My Red Hot Car" as it appears in (a) the warped version of the song, and (b) the original version of the song (Logic Pro X).

"I'm gon-na fuck you" in the last sentence is similar to "I'm gon-na" in the third sentence and "let me tell you girl" in the second, chopped up into a staccato rhythm that is punctuated by short signal dropouts (the fourth sentence generates the waveform depicted in figure 5.8). "With my" remains unaltered, while "red" is transformed into a percussive effect during which the start of the word is repeated numerous times at short intervals. The amplitudes before the peak of the attack of "hot" are deleted, making it almost sound like "hot." As elsewhere, the fourth sentence concludes with its last word, "cock," swallowed up by reverberation.

The song continues in this stuttering freeze-and-flow manner. While the synthesizer in the introduction warns the listener that this will be a glitch song, the cut-up vocal arrives somewhat unexpectedly and grabs much more attention than the cut-up synthesizer. The reason for this is that we tend to be especially sensitive to any manipulation of the *voice*, as Barry Truax points out: "The first sounds to which the ear is exposed as it develops in the fetus are human sounds, and from that point onward, the voice and human soundmaking are the sounds to which we are most sensitive as listeners" (Truax 2001, 33). Certain qualities and limitations are therefore associated with the vocal line in music, but "My Red Hot Car" upends them. Of course, a straightforward, unmediated performance of the lyrics would obviously be very different from the vocal constellation presented in "My Red Hot Car," which chops up the lyrics into a staccato musical form while removing the natural resonances of the sounds, as we saw above. When the

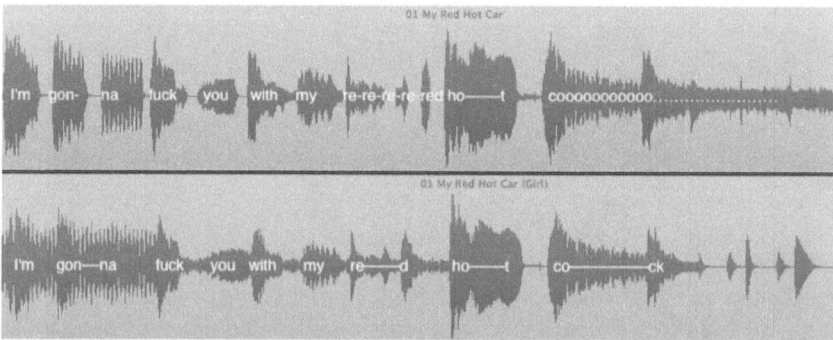

Figure 5.8
Waveform of the fourth sentence of Squarepusher's "My Red Hot Car" as it appears in (a) the warped version of the song, and (b) the original version of the song (Logic Pro X).

voice is manipulated to such a significant degree, we might assume that its role would change from bodily expression to purely musical sound. Yet our tendency to associate it with a human body that is delivering a message remains strong. The voice represents, first and foremost, an indexical sign of the human body and thus a clear path from source through musical performance to recording. In this way, the vocal in "My Red Hot Car" carries some of the same associations with a traditional, spatiotemporally coherent performance as "La Vida es Llena de Cables." And once again, we can discern two layers of music, the traditional and the manipulated, neither of which, in this precise context, makes sense without the other. It is the music's contradictory double meanings—it both is and is not a traditional performance; the glitches both are and are not part of the music—that supply its compelling tension.

Another reason why cut-up sounds seem to straddle the music's interior and exterior is that they are often entangled in associations of malfunctioning technology. As the sound of technological glitches is now frequently to be found in electronica music—and in more mainstream contemporary popular music as well—its shock effect has paled in comparison to the era when Knížák and Marclay used the sound of malfunctioning technology to their own radical aesthetic ends; as Kelly points out, "These sounds [of skips and stutters] are now simply another part of the sound palette of the digital producer" (Kelly 2009, 10). In their analysis of Madonna's "Don't Tell Me," Anne Danielsen and Arnt Maasø suggest that the track's cut-ups, which cause an acoustic guitar to freeze and flow in succession, might at first be mistaken for actual technological glitches, before the listener begins to reframe them as intentional musical gestures (Danielsen and Maasø 2009, 130–132). Although the sonic effects of "Don't Tell Me" are similar to those in "My Red Hot Car" and "La Vida es Llena de Cables," the latters' cut-ups are hardly likely to cause such confusion, since their rhythmic and cyclical nature is revealed right from the start. Moreover, since glitch sounds have become ubiquitous in music, they are now seldom mistaken for unintended "natural" glitches but rather are recognized as a musical effect.

While these glitchy skips, stutters, and signal dropouts might not be *mistaken* for actual technological glitches, they are not so easily released from their associations with technological failure. Malfunctioning sounds usually alarm us in relation to a problem or fault, and we most often react to them with frustration or unhappiness. Such sounds are hard to use subtly but instead tend to draw immediate attention to themselves. When these traditionally undesirable and certainly unmusical sounds are used to musical ends, they challenge our dichotomy between interior and exterior sounds, between the sounds that "belong" to the music and the sounds that reside beyond it.[14] Even if we understand the signal dropouts in "La Vida es Illena de Cables" and "My Red Hot Car" to be "missing" parts of the music, we want them to be part of the compositions (we know that they are meant to be there). Similarly, skips and stutters designed for aesthetic purposes are not what we traditionally think of as music either, yet they are somehow more artful and musical than glitches occurring naturally. This produces a further ambiguity, or perhaps a sense of double meaning: the skips and signal dropouts are at once unmusical elements (that are played with in a musical way) *and* musical elements in their own right.

We might clarify the notion of the glitch sound's double meaning by turning to Hutcheon's description of the double meaning that characterizes irony and parody. She recalls Ludwig Wittgenstein's metaphor of the duck–rabbit (similar to Janus's two faces/a vase)[15] to exemplify how one and the same thing might be experienced in two ways. Contrary to Ernst H. Gombrich (1969), who argues that we are *not* able to experience both readings of this metaphor simultaneously, Hutcheon suggests that, in the case of irony, we can: irony "implies a kind of simultaneous perception of more than one meaning … in order to create a third composite" (Hutcheon 1994, 59–60). She then refers to Stanley E. Fish (1983), who argues that the experience of ironic meaning involves a process in which the literal meaning is canceled out by the recognition of the "true meaning." According to Hutcheon, such assertions in fact limit the scope and impact of irony, because "this *either/or* theory does not account for the inclusive and simultaneous nature of ironic meaning" (ibid., 61, emphasis in the original). Instead, she suggests that, within an ironic frame, different meanings might be working *together* in order to create something new (ibid., 63). Similarly, some of the aesthetic power of malfunctioning sounds resides in their ability to be simultaneously experienced as the music's interior *and* exterior—our understanding of the glitching sounds as musical *coexists* with our understanding of them as sounds that disrupt the music. The meaning and function of glitch sounds inserted within a musical context are thus "both different [from that of actual glitches] and the same," to borrow Hutcheon's turn of phrase (Hutcheon 2006, 166). Put differently, sounds implying technological malfunctions challenge our traditional positioning of the border between the music's interior and exterior while at the same time reinforcing our need for one.

A Failure of Aesthetics or a New Compositional Palette?

Cut-ups and signal dropouts arguably draw immediate attention to themselves because of our associations of these sounds with malfunctioning technology, and because we often understand them to be the result of the schizophonic process of splitting the sounds from their temporally coherent origins. However, since glitch sounds have become ubiquitous in popular music in the digital era, partly because of the virtual cut-and-paste tool of DAWs, they are now seldom mistaken for unintended "natural" glitches and thus no longer generate a shock effect. Because of this process of normalization, Phil Thomson wonders whether glitch sounds have enough aesthetic power to survive. He suggests that the "aesthetics of failure," referring to the title of Kim Cascone's article from 2000 about glitch music, "might be on its way to being simply a failure of aesthetics" (Thomson 2004, 214). We, on the other hand, have argued that although glitch sounds such as skips and stutters no longer surprise us, they are still entangled in associations of malfunctioning technology, which creates a double meaning on several levels. The sounds both are and are not comprehended as musical sounds; they both constitute and disturb the music. Similarly, they both disrupt and sustain the music's flow, and they both strengthen and subvert the representation of a spatiotemporally coherent performance. By being experienced as disruptive in this way, they emphasize both the recording/production medium's ability to transparently mediate a message and its constant participation in that message, and this double meaning is part of the aesthetic quality of these cut-up sounds.

However, if the cut-up sounds' double meaning is not recognized—if they are instead experienced like any other sound, free from their association with malfunctioning technology—we would still maintain that their sound quality alone can be tremendously compelling and carry the day. Digital glitches, cut-ups, and dropouts have extended the music maker's compositional palette. For example, digital dropouts (in which the sound signal of one or more tracks is removed altogether) represent a new and unique musical element with a peculiar effect upon rhythm and sound.

While signal dropouts could be created in an analog medium as well through the insertion of leader tape—that is, blank, nonmagnetic tape normally used at the beginning and ending of a track—digital signal dropouts (which are created by simply *removing* the signal itself) consist of a characteristic digital silence that one cannot achieve in the analog medium. The fact that digital signal dropouts consist of complete silence makes them different from both traditional sounds and rests; signal dropouts are neither, and yet they contribute to the music with as much energy and force as either the sounds themselves or the traditional rests. The cut-and-paste tool's artful disruption of the acoustic qualities of sounds, in terms of removing part of a sound's attack and release/resonance,[16] demonstrates its potential impact on sound as well as rhythm. It also proves that *how* a sound starts or ends is as important as *when* it starts or ends (a musical aspect that cannot be captured by the traditional notational system and is often ignored in music analyses). Moreover, the traditional instrumentation that usually constitutes the foundation of the music can be replaced or supplemented by the sounds of skips, cuts, and stutters, which have the potential to produce new grooves with a unique sound. The use of the cut-and-paste tool in "La Vida es Illena de Cables" and "My Red Hot Car" produces a distinctive staccato effect and "partitioned" groove that manages to sound both disjointed and coherent at the same time. As listeners grow accustomed to the constant alternation between dropouts and sound fragments, the music begins to activate our musical expectations in this regard, and we come to expect sound to succeed silence, and vice versa. These expectations contribute to the persistence of a perceived forward movement in this otherwise halting groove; in fact, its apparent "lack of flow" constitutes a particular flow in and of itself—that is, the freeze and flow.[17]

The aesthetics of malfunction could, of course, be aligned with an ideological stance. Because digital technology allows for software programs that can be repaired or updated at will, and the digital sound represents completely high fidelity (see chapter 4), the consumer might have the impression that digital technology is near to perfection. Given this backdrop, glitch-inspired music could be interpreted as a self-reflexive critique of this assumed "perfection" of digital technology, or, as Phil Thomson puts it in his study of glitch music, as "work which inhabits the cracks in the digital dream" (Thomson 2004, 214). While these music makers may not always intend to construct such a critical commentary from within, glitch music's exposure of digital technology's potential malfunctions nevertheless contributes to removing the digital medium from its pedestal. Of course, the exposure of digital technology has an affirmative side as well, as Nicola Dibben has suggested in her compelling study of Björk's music: "The 'audibility' of digital technology ... can be understood as a positive embrace of the technological realm and its symbolism of the modern" (Dibben 2009, 80). There need not even be a contradiction between these respective interpretations. Of course, not everyone will appreciate the glitchy sounds of skips and stutters—some are simply unable to hear the minor malfunctions or spurious signals as musical gestures, or at least not as *appealing* musical gestures. While some will regard this aesthetics of failure as a failure of aesthetics, others recognize in these sounds the potential to become signifiers of artistic mastery, and the music that realizes them as a new frontier in the digital age.

6 Seasick Computers: Microrhythmic Manipulation in the Era of Endless Undo

Around the turn of the century, "seasick" grooves represented a sort of fad among contemporary R&B and hip-hop producers. They reached the mainstream on Brandy's innovative album *Full Moon* (Atlantic, 2002), produced by Rodney Jerkins, and perhaps even more so with Snoop Dogg's innovative album *R&G (Rhythm & Gangsta): The Masterpiece* (Geffen, 2004; in the following referred to as *Rhythm & Gangsta*), where several producers, among them J. R. Rotem and Josef Leimberg, contribute their takes on the trend. Common to these seasick grooves is the fact that their "feel" aspect is almost overdone, leading to what Anne Danielsen elsewhere has labeled the "exaggerated rhythmic expressivity of the machine" (Danielsen 2010a, 1). The microrhythmic excess of such grooves represents in many ways the antithesis of the clarity in sound and precision in timing that characterized the early days of digital music processing in the 1980s (see, e.g., chapters 2 and 3 in this volume). Whereas Prince's "Kiss" recalls late 1970s disco funk, this new musical trend in one sense resembles the "deep" funk grooves of 1970s bands such as Parliament.[1] However, this onetime "organic" feel has now been given a distinctive computerized "twist" that testifies to the presence of the manipulative use of digital technology. Through the exploitation of the new possibilities for microtemporal editing brought about by digital recording, the temporal discrepancies between the different rhythmic layers forming the repeated pattern of the groove simply transgress the boundaries inherent in the perceptual capacities of a human musician. There is a limit to how far "out" a human musician can place his or her own beat without abandoning the firm ground that is needed to maintain the steadiness of the pattern that constitutes the groove.

We will start the narrative here by recapitulating the extraordinary opportunities for controlling and manipulating the temporal axis of music that were brought about by digital recording. Then we will proceed to the analyses of two songs from Snoop Dogg's *Rhythm & Gangsta* where these new opportunities have made an audible mark on the sounding result.

Taking Control in the Temporal Domain

Musicians have always displayed a compelling ability to control the temporal domain. Prior to digital sound editing, musicians who were responsible for the groove had to develop this skill in order to survive in the business.[2] When digital drum machines with "natural" drum samples came into use in the early 1980s, many thought that the era of live drummers in professional studio recording had come to an end. However, as discussed in chapter 3, there were still several problems with getting the machines to "groove." One was the absence of dynamic development in the sounds they produced, both within a given sound and from one sound to the next. While sound quality increased significantly with digital drum sampling, which allowed for a new sonic richness in sequencer-based grooves, it was still very difficult to produce the variation in sound that was typical of (and enjoyable about) played grooves (Inglis 1999). Another limitation to sequencer-generated grooves in the first years of sampling and MIDI was a lack of microtemporal flexibility. The opportunities for manipulating the temporal dimension were rather limited and consisted by and large of either quantization or the so-called humanizing function. The former simply confirmed the machinelike feel of sequencer-based grooves rather than challenging it. The latter did indeed "add" microtiming to programmed events, but this variation was random, in contrast to most variation produced by skilled musicians. Nevertheless, it compensated to an extent for the stiffness of sequenced rhythm, which proved to be a persistent association with sequencer-generated grooves into the 1990s (Inglis 1999).

In many genres within the field of electronic dance music (EDM) to the present day, in fact, this machinelike timing remains a distinguishing stylistic feature and even a preference well after other alternatives to it became available in the early 1990s (Zeiner-Henriksen 2010b). In most other groove-based genres, however, it has been an important aesthetic challenge to develop sophisticated microrhythmic designs focused on temporal relationships or microtiming. A first step in making this possible even in sequencer-based grooves was to combine the sampling of longer stretches of music—that is, one or more bars—with sequencing through the implementation of audio sequencing in traditional MIDI sequencer programs. This was, however, not common practice before the early 1990s, when sequencer software programs, such as Steinberg's Cubase and Emagic's Logic MIDI, were finally equipped with audio sequencing capabilities (see chapter 1).[3]

A second step in the creation of microrhythmic designs was sound editing. This process started with the development of sample editors that would prepare samples for sample-based instruments such as the Fairlight CMI and the Synclavier. Connecting a sampler to a personal computer (a Macintosh or Atari) added a larger screen and extra computing power to speed up the editing. ProTools, for example, began as a Mac-based sample editor, Digidesign's Sound Designer. This software morphed into Sound Tools, a direct-to-disk recording system launched in 1989 that allowed for nondestructive editing in stereo and also included some simple digital processing. In 1991 a four-track version called Pro Tools was released, and shortly thereafter, Digidesign's Digital Audio Engine became available to other manufacturers, opening up Pro Tools' hardware to users of sequencer software (Burgess 2014, 145; Hofer 2013).

Sound editing, of course, did not arrive with digital recording; editing and even cut-and-paste were in use already in the analog era. First, as Richard J. Burgess (2014) points out, arrangers, composers, and orchestrators have always been able to cut and paste by applying a few strokes of the pencil to the written score, or through a verbal instruction to the musicians, in this way moving whole sections or smaller fragments around as needed (Burgess 2014, 137–138). Second, as was made clear in the previous chapter, magnetic tape allowed for cut-and-paste operations on recorded sound as well. For example, sections of music could be copied or recorded on a separate tape machine and then reinserted at the desired location on the master tape using the "fly-in" technique.[4] However, as Burgess points out, while magnetic tape allowed producers to cut sound into pieces and reassemble it, "digitization breaks sound into microscopic pieces that are reassemble-able in almost any configuration" (ibid., 137). One of the radical shifts of digital technology, that is, was precisely the extent to which one could cut sound into pieces and reassemble it. Digital sound editing is so simple and fast that there are almost no limitations to how many edits can be done. Thus, whereas sound editing in the analog era was used first and foremost to improve the overall form of the song (a chorus could be doubled or a bridge removed) or to correct serious mistakes, digital sound editing has become a tool for *perfecting* a recording, as well as for manipulating and experimenting with the design of the music: a snare drum can be moved thirty milliseconds earlier in time, either to correct a misplaced stroke or to add some extra push to the groove; a lead vocal or a guitar solo can be compiled out of several takes, and the feel of that lead or solo can be changed by moving a phrase—or parts of a phrase—earlier or later in time. Digital sound editing, in other words, is yet another example of the fact that many digital developments represent, first and foremost, a quantitative leap in the ability to perform operations that were in principle also possible with previous technology. However, as is clear from both the microrhythmic manipulation discussed in this chapter and the glitch aesthetic dealt with in the previous chapter, this quantitative leap is sometimes so significant that the result is better described as a qualitative, sometimes even historic, change.

For a long time the two lines of development described above—the integration of longer loops of audio in the user-friendly MIDI sequencer, and hard-disk recording, with its extended opportunities for sound editing—were more or less distinct. Accordingly, what might be regarded as prototypes of digital audio workstations (DAWs) tended to focus on either sequencing (Steinberg's Cubase and Emagic's Creator, Notator, and Logic Series) or recording (Digidesign's Sound Tools and, later, Pro Tools). The graphical interface of the Pro Tools software, for example, was modeled after analog tape-based recording studios, whereas the interface of the sequencer-oriented software was an extension of earlier MIDI-sequencing software. When these two lines of development eventually came together in the early 1990s, a new world of opportunity opened up for manipulating the time line of multitrack sequencer-based grooves. The next important step in this regard came in 1996, when Steinberg introduced Cubase VST, which could record and play back up to thirty-two tracks of digital audio on a Mac (Musicradar 2011). The software offered a tapelike interface for recording and editing, as well as the entire mixing desk and effects rack that were common in analog studios. Its most revolutionary aspect, however, was that all of these operations could be done in the software alone (no external hardware required).

The possibilities for manipulation, and thus sheer power, of the contemporary DAW are extreme and the happy result of many disparate developments. Among other things, it provides a wide palette of digital signal-processing resources (as discussed in chapters 2 and 7). Here we will focus on music that was fundamentally marked by the integration of digital recording—with its endless editing and processing possibilities—into traditional sequencer software. The sequencer provided easy multitrack looping, a feature that was held in high esteem in the creation of repetitive multilayered groove-based music, and hard disk recording facilitated nondestructive editing at all levels, allowing for more "custom-made" designs of both the basic pattern of the groove and its overall form. In combination with the marvelous "undo" function, this flexibility led to a new practice of "endless" editing across all genres of popular music.

Snoop's Muddy Microrhythms

Around the turn of the millennium, some years after the world of sequencing and the hi-fi world of digital recording had begun to converge, the potential for manipulating the microrhythmic design of recorded musical events was starting to make its mark on African-American-derived groove-based music. It is no surprise that genres related to hip hop were pioneering in this respect. Hip hop was already tightly coupled to developments within music technology, such as the sampler and the sequencer, and the rap aesthetics at the microrhythmic level also invite experimentation with new and more extreme rhythmic feels. Laidback timing in the rap part of a hip-hop production is common and evokes the timing of the solo parts in other groove-based genres. (In jazz, for example, both the lead vocal or lead instrument and improvising solo instruments often display a very free and laidback sense of timing, as if they were floating atop the groove.) What was new with digital audio sequencing and editing was that huge timing discrepancies between rhythmic layers could become part of the groove itself, almost as though there were different layers in the accompanying tracks that did not belong to the same timing reference. Perceptually, this is sometimes experienced as if multiple pulse references are at work. The accomplishments of the Soulquarians collective toward the end of the 1990s—the most well-known member is probably Erykah Badu—were pioneering in this respect. Inspired by J Dilla, these artists started to insert marked microtemporal discrepancies between the basic rhythmic layers, producing a series of grooves with a peculiar, "seasick" rhythmic feel.

Brandy's single "What about Us" from the album *Full Moon* (Atlantic, 2002) is an early mainstream example of how these new opportunities could produce microtemporal gaps between rhythmic layers in the postproduction phase.[5] Thus, when Snoop released the *Rhythm & Gangsta* album in 2004, extremities in machine-aided microrhythmic design were not new. However, the extent to which this capacity is realized on this album is nonetheless surprising—most notable are the experimental grooves of "Bang Out" (track 2), "Can I Get a Flicc Witchu" (track 4), and "Fresh Pair of Panties On" (track 12). Interestingly, these three songs have different producers. "Bang Out" is an early accomplishment of J. R. Rotem, a jazz pianist who started out as a hip-hop producer but later turned to production within contemporary R&B and pop. "Can I Get a Flicc Witchu," which features P-Funk's Bootsy Collins on backing vocal, was produced by Josef Leimberg, and "Fresh Pair of Panties On" was produced by Ole Folks. This indicates that at the time of the release of *Rhythm & Gangsta*, the seasick feel produced by the insertion of extreme discrepancies into the rhythmic foundation had become a trend, thanks to the ease of use of the compositional tools provided by the DAW. In the following, we will analyze two of the more extreme machine-generated microrhythms on the album, from "Can I Get a Flicc Witchu" and "Bang Out," respectively.

The groove in "Can I Get a Flicc Witchu" consists of a programmed bass riff and a drum kit, along with vocals that are mainly rapped. The texture of the groove is simple and open, though its muddy microrhythmic relationships present as complex and almost chaotic. Its basic unit is a two-bar pattern. In an earlier analysis of the tune, Carlsen (2007) proposed the hi-hat pattern on quarter notes as the constitutive reference for its basic pulse (see figure 6.1). This reference, however, is not very precise at a microtemporal level, particularly because the "hi-hat" sound on the quarter notes is not very hi-hat-like—it sounds more like a combination of a deep breath and an open hi-hat played in a careless manner. This beat, then, is in fact a beat bin: it has considerable extension in time and no clear beginning or end, at least compared to the sharp attack of a normal hi-hat, meaning that the tolerance for multiple and/or vague rhythmic events at this groove's beat positions is quite high. The sampled vocal and the remainder of the drum kit relate to this metric reference,[6] despite the fact that it is also not isochronous. The length of the beats is gradually shortened, so that beat 2 is shorter than beat 1, beat 3 shorter than beat 2, and so on (see figure 6.1). This may

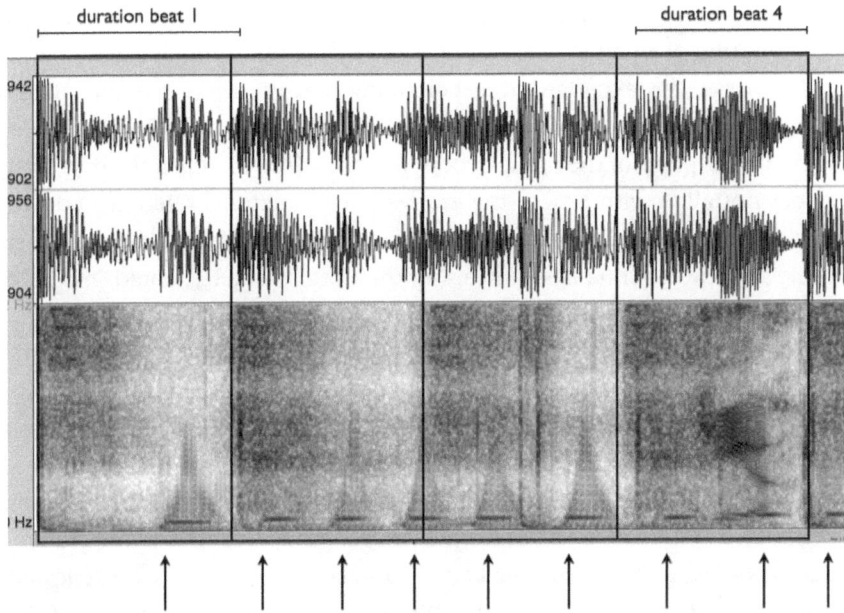

Figure 6.1
Waveform (amplitude/time) and spectrogram (0–7,000 Hz) of the first bar of "Can I Get a Flicc Witchu" (Praat, version 5.1.03). Isochronous grid of quarter notes = black lines. Bass riff onsets indicated by arrows.

be the result of the use of tempo automation, a function that was available in the DAW at the time of production of *Rhythm & Gangsta*. In this case, it appears that the tempo has been set to gradually increase from approximately 92 bpm (one quarter note = approx. 652 ms) to approximately 98 bpm (one quarter note = 612 ms) over the course of one bar (in 4/4 meter). This manipulation also contributes to the feeling that each pulse is a wide beat bin and a general vagueness as to the positioning of rhythmic events. Consequently, the doubled hand claps at beats 2 and 4 are easily contained by the beat bin, as is the sampled vocal.

Despite this loose metric framework, the deviating bass pattern is nevertheless striking. It follows its own peculiar schematic organization (see the arrows in figure 6.1) and is a main reason for the seasick rhythmic feel of the tune, because it is not related to the 4/4 meter and does not conform to a regular periodicity of its own either. It could be heard as an inaccurate and unstable variant of a triplet pattern commencing on beat 2, but as it gradually lags behind this metrical reference, the listener might begin to experience it as some kind of quintuple pattern (five against two),[7] or even as "free rhythm"— a pattern that is not related to a metric grid at all.

The bass riff is experienced as belonging to a rhythmic scheme that is completely different from the other rhythmic layers in the groove (perhaps with the exception of the rapped vocal, to which we will return shortly), and it does not align with any of the groove's other metrical reference points. Such a free-floating rhythmic feel is very difficult to play live, because most musicians, in the interests of keeping the pattern stable, would gravitate toward a shared metrical reference.

This peculiar feel has most likely been produced in Pro Tools post recording,[8] either by adjusting the temporal onsets of the programmed events forming the bass riff pattern until the sought-after seasick effect was achieved or by recording the bass riff separately, in free rhythm, or sampling it from a different source. In the latter two cases, when the producer attempts to integrate such a sample or recording into the main groove, it must be warped—twisted or bent to match the length of the repeated unit in the groove, often by temporally stretching or shrinking parts of the sample (or the entire thing). The producer could also cut out a piece of the source (a recording or a sample) that has the exact length of the loop and paste it into the new musical context, regardless of possible mismatches in meter and tempo. If the tempo or meter of the source is different from the target groove, the rhythmic structure will be "perceptually warped" as well, in relation to its new context. If it is also isochronous, the periodicity of the sample will likely be disturbed, since the first or last beat of the sample/recording is likely to be too short. Depending on the degree of the mismatch, this will result in a more or less dramatically halting feel when the sample is looped.[9]

The bass riff of "Can I Get a Flicc Witchu" is an example of a rather extreme temporal/rhythmic mismatch between the main groove and a source sample/recording—so extreme that it becomes the groove's signature element, coloring one's whole impression of the song. Snoop Dogg himself picks up the seasick feel in his rapped lead vocal in the first verse of the song, which partly relies on the bass loop's alternate metrical framework. The rap starts out with a rhythmic pattern that relates loosely to the eighth-note level of the 4/4 meter, but gradually it becomes more and more influenced by the rhythm of the bass. This is particularly striking in the second sentence of the rapped verse. Here, the natural speaking rhythm of the sentence suggests a beat-oriented pattern with one syllable on each eighth note (except for the double sixteenths on 2-and) in the following manner: "Rock"—"this"—"beat"—"'cause you"—"are"—"so"—"fresh." However, as can be seen in the spectogram in figure 6.2, the syllables are not at all aligned with the grid of the eighth-note pulse. In relation to this pulse, in fact, the timing of the rap vocal comes forward as extremely late—so late that it almost ends up on off-beats. An alternate and perhaps more revealing perspective, then, would be to regard the rapped vocal as relating primarily to the pulse of the bass riff: the syllables either lead up to ("Rock"), are centered around ("this," "'cause"), or simply begin with ("beat," "are," "so," "fresh") the onsets of this alternate layer of "beats." Perhaps the most accurate interpretation is indeed to hear the rap as influenced by *both* of these reference structures—that is, by the eighth notes of the straight 4/4 meter and by the bass riff's highly nonisochronous, alternative pulse. In any case, the feel of this rap would have been much more difficult to achieve *without* the accurate inaccuracy of the bass loop, making this song an excellent example of the way in which a new gestural language produced by a machine can be mimicked to great aesthetic effect by a human performer.

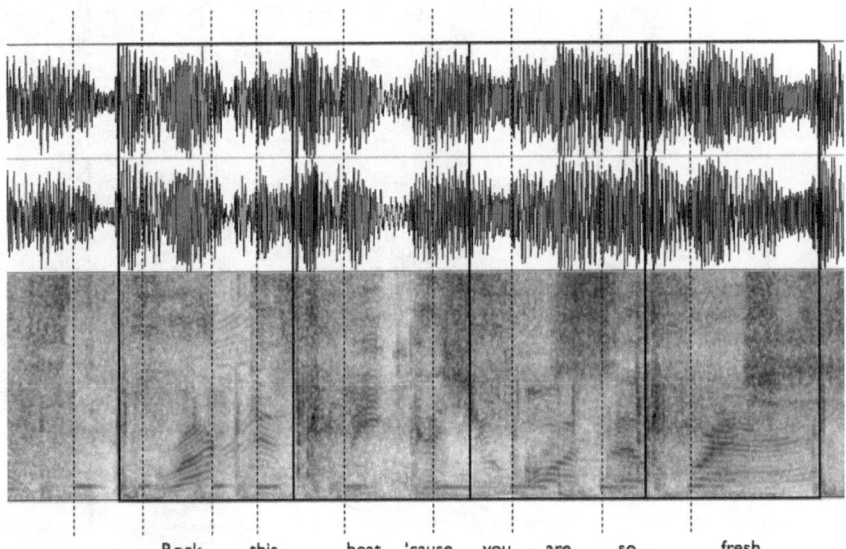

Figure 6.2
Waveform (amplitude/time) and spectrogram (0–7,000 Hz) of bar 10 of "Can I Get a Flicc Witchu" (Praat, version 5.1.03). Isochronous grid of quarter notes = black lines. Bass riff onsets = stippled lines.

Similarly, "Bang Out" is characterized by a peculiar machine-generated rhythm, and its seasick feel arises from the clash between two alternative subdivisions in the groove. The piano plays straight subdivision on all sixteenths, supported by a hi-hat on the quarter notes, the backing vocal line ("Bang out …"), and the synth pick-up to beat 3. At the same time, the combined bass and bass drum attacks imply a triplet-related pattern (see figure 6.3).

The triplets in bass and bass drum are not clearly defined and feel rushed in relation to the piano pattern. When mapped against an isochronous grid of triplets, it becomes obvious that the whole pattern is five milliseconds early in relation to the basic pulse articulated by the hi-hat and backing vocals. Most likely, the bass drum and bass layers have been displaced—that is, moved earlier in time—in the postproduction process. In addition, the piano pattern appears to have been moved approximately 10 milliseconds *later* in time in relation to the backing vocal/hi-hat reference. The discrepancy between the bass/bass drum layers and the piano pattern thus totals a noticeable fifteen milliseconds. In a song like "Can I Get a Flicc Witchu," small displacements like these would probably not have been noticed at all

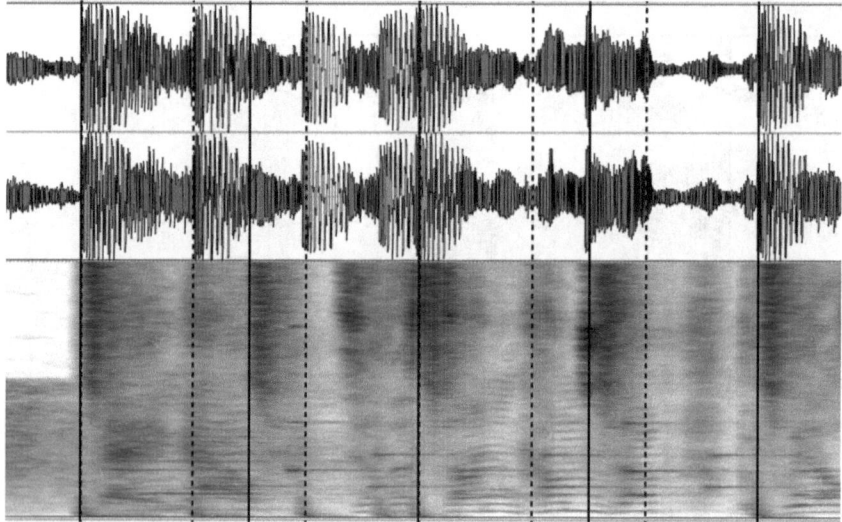

Figure 6.3
Waveform (amplitude/time) and spectrogram (0–10,000 Hz) of "Bang Out," bar 1 (Praat, version 5.1.03). Grid on quarter notes = black lines; grid on triplets (3:2) = stippled black lines.

due to the chaotic microtiming, blurry sound, and many extended beats. In "Bang Out," however, the beat bin—the perceived temporal extension of a beat according to the musical context[10]—is considerably narrower, because the sharp attack of the piano defines a more accurate position for the beat and the bass and bass drum sound is also more focused, with a clear core of energy. This results in a lower tolerance for temporal deviations (or "rhythmic tolerance" [Johansson 2010]) in "Bang Out" than in "Can I Get a Flicc Witchu."

Clashes between different metrical grids at the level of subdivision represent a feature that appears in various groove-directed musics, both played and programmed.[11] In played music, this feat depends on (and is circumscribed by) the skills of the performers, who will typically be taxed simply to keep two clashing patterns of subdivision stable in such a musical context. In a computer-based groove like "Bang Out," where almost all of the rhythmic layers are looped and controlled by a sequencer, the task is more straightforward: one can experiment with potentially conflicting patterns and even vary the microtiming in one or all of them without having to account for the perceptual and performative limitations of human beings. The bass and bass drum pattern in either "Bang Out" or "Can I Get a Flicc Witchu" is almost unplayable, at least repeatedly, and testifies to experimental use of the new opportunities for microrhythmic manipulation enabled by the integration of digital sound editing and MIDI and audio sequencing in the DAW.

Manufacturing the Inner Dynamics of a Groove

The DAW presented new opportunities for optimizing and experimenting with the microrhythmic design of grooves. Nondestructive editing was a particular virtue in this regard: when one can return to previous stages in the creative process by simply clicking "undo," one can take more risks and can test untraditional or extreme solutions. As we can hear in Snoop Dogg's *Rhythm & Gangsta* tracks, the music has been warped into new rhythmic feels that could not have been achieved by musicians or any preceding technological tools. These feels derive specifically from the DAW's capacity for taking control of and manipulating events along the temporal axis.

This enhanced control at the level of microrhythm made the DAW particularly attractive within genres such as hip hop, neo-soul, and contemporary R&B, all of which are governed by a groove-based aesthetics that values perfecting the rhythmic design of the basic unit of the given groove. This unit, which is repeated seemingly endlessly throughout the song, is generally very limited in duration but not in its impact, and it demands the utmost attention. For a groove to succeed as a groove, it has to be designed so that it engages listeners in a coproductive, bodily process. Rather than experiencing the groove from some arbitrary analytical distance, the listener should become engaged in the unfolding of the rhythm from within—that is, the rhythm is almost coproduced by the listener as he or she moves along with it, virtually (in the head) or actually (through dancing). Along these lines, Gilles Deleuze finds it necessary to distinguish between two types of repetition—dynamic and static—which correspond to repetition as experienced from positions inside and outside the given process, respectively. The latter concerns only the overall effect: it *results from* the work and refers back to a single concept that is identical for all repetitions of the basic pattern. The former concerns the acting or productive cause, an internal difference that drives the process onward and that is incorporated in the repetition. According to this interior perspective, the groove is not shaped at once but rather comes into being like the "evolution" of a bodily movement—or, in other words, like a gesture. Deleuze writes, "In the dynamic order there is no representative concept, nor any figure represented in a pre-existing space. There is an Idea, and a pure dynamism which creates a corresponding space" (Deleuze 1994, 20; see also the discussion in Danielsen 2006, chap. 8).

In order to keep the listener engaged in the groove in this way, the inner dynamics of the groove's basic unit are crucial. To do so, the artist can introduce a compelling interaction between different layers of rhythm at a *structural level*, as was the case with all sequencer-based grooves prior to the microrhythmic manipulation of rhythmic events enabled by the DAW. This type of rhythmic interaction probably gave rise to the on-the-grid aesthetics of electronic dance music: all of the events in the groove had to be on the grid because there were no other options. Driving such grooves forward could only be done through structural tension between the basic rhythmic figures of the groove—for example, by constructing a polyrhythmic fabric of rhythms and counterrhythms. A common example would be the 4:3 figure, where four isochronous strokes on, for example, a synthesizer are juxtaposed against three beats of the basic pulse.[12]

However, as illustrated by Prince's "Kiss" in chapter 4, musical sounds in and of themselves can also generate inner dynamics that invite movement. Probably because of the limitations of microtemporal control in machine music prior to the DAW, electronic dance music producers became extremely concerned with picking and shaping the right sound for each rhythmic event, as well as with producing momentum through such effects as the opening up of hi-cut filters and the like. Zeiner-Henriksen (2010a) also shows how bass drum sounds in analog drum machines, in particular the Roland TR-808 and TR-909, were manipulated to enhance the drive of EDM grooves at the microlevel. By tweaking the settings on the drum machine, the core, or perceived heaviness, of the sound could be displaced a little bit later in time (the onset of the sound was still controlled by the grid of the sequencer), adding to the drive of the groove.

The most common means of driving a groove forward prior to the DAW, however, was to allow the musicians to work out the optimal microtemporal relationships among their parts. When digital audio editing became available and integrated into sequencers, of course, this exact process could suddenly be done on the computer by moving events back and forth on the temporal axis. Whether human or machine in origin, interesting microrhythmic designs often consist of rhythmic events that are slightly askew of their expected temporal positions according to the song's structural scheme, or of rhythmic events that are located in altogether ambivalent temporal positions.[13] In the grooves analyzed in this chapter, events that undercut the expected schemes succeed to such an extent that they introduce viable alternative schemes instead. In "Bang Out," the seasick feel is created by a clash between different matrices of subdivision, which is very easy to accomplish in a sequencer. This is combined with the manipulation of the positioning of audio tracks, which have most likely been moved back and forth on the temporal axis in the sequencer by applying a delay or a predelay to an entire track until the wanted effect—what might be described as the characteristic "stretched" beats—was achieved. The use of machines ensures that the discrepancies in the groove are steady, an effect that is difficult to create with humans, at least until they have motorically mapped the new feel. In "Can I Get a Flicc Witchu," the bass riff is so off in relation to the main metric grid that it almost threatens the stability of the groove. Because its irregular structure also makes it hard to remember, it ultimately contributes a series of moments of perceptual confusion. It never becomes redundant but instead seems to reestablish itself each time it reappears. The instability produced by such deviating patterns is very effective in enhancing the inner dynamics of the groove and keeping the listener engaged in understanding it.[14] When the song's rhythmic structure never settles, the listener lacks firm ground—that is, the potential regularity of the pattern is subdued almost at the moment that it emerges. When used to an extreme—and in particular when produced by machine-generated rhythms, where there are no human limitations imposed on the process—these deviations, taken together, may in fact transcend the limits of perception, with the consequence that the whole rhythm falls apart. Even though the machine has no limitations as to what it can produce, there are still limitations on one's ability to understand the pattern *as a pattern*. The effect of such extreme microrhythmic designs, then, is profoundly dependent upon the listener's training and familiarity with the musical style in question, as well as the cultural and historical context. Balancing on the edge of structural dissolution must have been a real challenge for a commercial artist like Snoop Dogg when he was recording the most extreme tracks on the *Rhythm & Gangsta* album.

A New Gestural Language

Visual editing of digital audio allowed for an unprecedented level of rhythmic sophistication. Even though rhythmic sophistication was of course also possible to achieve with musicians, it would usually require many hours of practice to get exactly the right feel. For a producer aided by visual digital editing, it takes only seconds to alter the timing of one or more events in a repeated rhythmic pattern. In the end, this quantitative increase in the capacity to complicate rhythms resulted in a qualitative change in the music, because so much more could be done during the time available. Moreover, the ability to "undo" encouraged extreme experimentation and, in turn, those entirely new rhythmic feels that have become signatures of this era in popular music history. As we have seen, the DAW represented a very powerful tool for generating and optimizing new, as-yet-unheard rhythmic feels and also allowed for a new complexity in the structural and microtemporal features of computer-generated groove-based music.

As with many of the other digital tools discussed in this book, the DAW could be used to either perfect an existing groove aesthetic or generate something completely new. The former can be described as *machine-aided* perfection of a groove, and the latter as *machine-generated* rhythmic feels (because they could not have been achieved by any other means). As demonstrated in the analyses above, both "Can I Get a Flicc Witchu" and "Bang Out" are fundamentally characterized by their exploitation of the DAW's new opportunities for manipulating and generating microrhythmic feels. The use of clashing subdivisions, a main ingredient in "Bang Out," and the warped sample-produced mismatch between the main groove and the bass riff in "Can I Get a Flicc Witchu" could be taken to a new level with the DAW.

As noticed above, the seasick feel produced by the warped sample in "Can I Get a Flicc Witchu" is picked up by Snoop Dogg's rapped lead vocal. In "Bang Out," as well, the vocal bears traces of the typical phrasing that often arises atop clashing subdivisions, audible as a distinct in-between shuffling of the subdivisions of the beat, as though the vocal were drifting between duple and triple subdivisions.[15] Both of these adjusted vocal alignments are examples of how people can be inspired by and learn from the machine.[16] The new rhythmic feels produced by the DAW at the turn of the millennium thus represent a new chapter in the merging between human and machine that digital technology has brought about in the field of music. They clearly demonstrate the ways in which new gestural expressions generated by digital music technology can be picked up by people and made a part of a human gestural vocabulary. This illustrates the two-way interaction between musicians and technology. The story of digital technology in the field of music is not simply about people trying to get the machine to mimic the human world. It is also about people wanting to play and, as we will discuss in the next chapter, even sing like the machine.

7 Autotuned Voices: Alienation and "Brokenhearted Androids"

We regard the human voice as the most *human* of all instruments, to the extent that even in the field of pop music—where masks are the rule rather than the exception, and where people tend to expect that most parts of the musical outcome have been through heavy processing by various sound-production technologies—obvious manipulation of the human voice still sparks controversy.[1] And of all of the things done to the human voice in pop production, autotuning—that is, digital pitch correction—appears to be the most controversial. A peak in the heated debate on the use of this digital signal-processing tool is the use of Auto-Tune in the 2010 version of the British reality show *X Factor*: "A major scandal is brewing over revelations that the show enhanced the vocals of some contestants during Saturday's season premiere, with some of the very obviously tweaked vocals so clumsily edited they could have been included on a T-Pain single" (Kaufman 2010). The scandal arises from the audience's feeling of being fooled—no small thing where the voice is concerned—a sentiment contained in *Daily Mirror* journalist Tom Bryant's statement about contestants and fans having been "cheated" by "audio wizardry" (Bryant 2010). Even the artists taking part in *X Factor* expressed their deep concern. Former 2008 finalist Austin Drage stated: "The moment you start putting effects on the voice you lose the raw sound. ... It is an unfair advantage and it is like runners taking drugs to run faster" (quoted in Bryant 2010). Daniel Evans, a former contestant who reached the live finals in 2007, described Auto-Tune as if it were a disease: "I am big campaigner against Auto-Tune in the music industry. ... It's like an epidemic. ... Sad times for real music fans" (quoted in Kaufman 2010).

Why does the digital manipulation of the voice elicit such strong reactions? We will return to this question toward the end of this chapter, once we have presented the technological principles behind Auto-Tune and reviewed milestones in the history of the ultimate example of a technology that has clearly become subject to unanticipated applications.

Use and "Abuse": Pitch Correction as Vocal Effect

Autotuning technology demonstrates—perhaps better than any of the other new digital tools discussed in this book—how new tools are used very differently in different musical contexts. Despite harsh criticism and the attendant controversies, digital pitch correction, like the microrhythmic "correction" techniques discussed in the previous chapter, has become a valuable source of completely new sonic effects in vocal production. The first digital pitch-correction software was invented by Andy Hildebrand, a research scientist in the geophysical industry who worked to develop digital signal-processing methods that could be used to identify oil deposits through the analysis of sound reflections from subsurface layers of the Earth. In 1979 he left the oil business to study music and realized that the method he had developed to analyze seismic data could also be used to analyze and correct pitch in audio signals. Hildebrand's digital pitch-correction plug-in, Auto-Tune, was made commercially available through Antares in 1997 (Antares Audio Technology 2015).

Auto-Tune is the digital age's answer to the analog Vocoder, which first appeared in several popular music recordings during the 1970s and is famously associated with Kraftwerk and the Alan Parsons Project, though artists as diverse as Pink Floyd, Jean Michel Jarre, Daft Punk, and Madonna have also used it. The Vocoder is an electronic musical instrument that analyzes and transforms the frequency content pattern of an input signal (usually a voice) into electronic information. This information is then used to control another input (carrier) signal of the Vocoder (usually a synthesizer). If one is using speech as the input signal, the spectrum of the speech sounds controls a multiband filter that modulates instrument pitches, superimposing "a replica of the voice's energy patterns on to the sound of the instrument" (Anderton 2006).[2] The sonic outcome is a re-created synthetic version of the analyzed input signal (that is, a synthesized voice). Whereas the Vocoder is an analog synthesis procedure, the Auto-Tune plug-in is based on digital signal processing of the numeric representation of the sound wave. Auto-Tune identifies the dominating periodic frequencies, or pitched notes, in the signal using autocorrelation techniques and adjusts them to the nearest periodicity corresponding to one of the notes in a predetermined scale. In short, it changes the pitch of the signal while keeping its other features intact. In addition, while the sonic result of using a Vocoder on a vocal sound is a hybrid between the voice and a synthesizer, the sonic result of using Auto-Tune on a vocal is a pure vocal sound, but one deprived of typical human characteristics, such as vibration or sliding transitions between tones.

It was soon discovered that the new sounds and procedures provided by Auto-Tune could be used in a more experimental way to create fascinating new sounds. The first and perhaps still best-known instance of the opaque use of digital pitch correction is Cher's "Believe," released in December 1998. Producers Mark Taylor and Brian Rawling almost stumbled over the effect as they experimented with the retune setting in Auto-Tune, a crucial parameter in the software that decides how long it will be before a singer's voice is adjusted to the "correct" pitch. The retune speed can be set anywhere between zero and several hundred milliseconds. Zero means that the software finds the nearest note and changes the output pitch instantaneously, which makes the singer's voice sound jumpy and mechanized, because it entirely eliminates the natural transitions between notes. Cher's "Believe" became a big hit, and in an interview given a few months after its release, the producers tried to conceal their new discovery, claiming that the effect was produced by the newly released Digitech Talker Vocoder pedal (Sillitoe 1999).[3] Commenting on the Cher effect, Auto-Tune inventor Hildebrand later said, "I never figured anyone in their right mind would want to do that" (Tyrangiel 2005).

The Cher effect recast autotuning as more than a means of "cheating" the listener, and its roster of potential sonic effects reached into new genres when R&B artist/rapper T-Pain used pitch-correction software to process his lead vocal on several tracks on the album *Rappa Ternt Sanga* (Jive) in 2006. In "Believe" the robotic effect had been reserved for selected phrases and intentionally juxtaposed with Cher's "normal" (or transparently mediated) voice. With T-Pain the effect suddenly supplied a characteristic aspect of the entire lead vocal, which sounds like a mixture of a man and a sad machine. In this case, as well, pitch-correction software was probably also used to generate the background vocals, which sound like pitch-shifted copies of the lead vocal. Digitally generated background vocals were one of the many creative features offered by Auto-Tune's competitor Melodyne (Celemony), released in 2001. In addition to pitch correction and pitch-shifted replications of the lead vocal, Melodyne had facilities for time stretching and melody rebuilding and offered far more creative possibilities than the early versions of Auto-Tune (Johnson and Poyser 2001).[4]

The creative, opaque use of pitch correction tools flourished in the 2000s and has become associated with certain topoi in popular music. When using autotuning for creative purposes, artists usually engage with it as an input device—that is, the artist listens to the sound produced by the effect while singing. (When used to correct pitch, on the other hand, artists and producers engage with it as an *output* device—that is, it is applied to the vocal after recording.)[5] In hip hop, digital pitch correction is often used in a manner akin to T-Pain's instincts, to express alienation, sadness, or loss. Kanye West's album *808s and Heartbreak* (Roc-A-Fella Records, 2008) is a classic effort in this respect, capturing, according to the *Washington Post*'s review, "the isolation, paranoia and longing of 21st-century city life" (Richards 2008). The discourse surrounding Kanye's release also illustrates well the win-win situation brought about by digital pitch-correction tools. The act of correcting or "cheating"—making singers out of people who cannot sing—and the act of creating are intimately mingled when one employs this tool. Auto-Tune clearly assists Kanye in satisfying the responsibilities of lead vocalist (with perfect intonation) on a professional recording. In an interview at musicradar.com, he says: "If I sing off-key, it [Auto-Tune] really points that out. It points out the bad notes. So what I have to do is to sing more perfect" (Rogerson 2008b). At the same time, autotuning enables a particular sort of vocal expressiveness that is beyond the reach of human singing. West's obvious indulgence in Auto-Tune is clearly also rooted in a deep fascination with this signature sound.[6] The sad, mechanistic sound of his autotuned voice suited the overall theme of his album, which centers around emotional distance, loneliness, and heartbreak, probably inspired by turbulent events of his own life. In a review of the album, *Rolling Stone* music critic Jody Rosen concludes, "Kanye can't really sing in the classic sense, but he's not trying to. ... Kanye's digitized vocals are the sound of a man so stupefied by grief, he's become less than human" (Rosen 2008). Auto-Tune has also been used to describe or evoke other conditions characterized by alienation from one's emotions or physical sensations, for example as a consequence of drug use. This is the theme of Frank Ocean's

"Novacane" (Def Jam, 2011), for example, where the voice is literally auto-tuned but the process is also used in the lyrics as a metaphor for the protagonist's numbness, which feels like a "pitch-corrected computer emotion."

Auto-Tune's connotations of the robotic and nonhuman have also been used to disrupt stereotypical notions of race and/or gender, particularly around the reception of female artists within electro-pop and black R&B. The sound is often coupled with imagery depicting exaggerated femininity and/or hyperembodiment—that is, a body that comes forward as either perfect in and of itself or otherwise cultivated beyond the human. In her essay on robo-divas in contemporary R&B, Robin James (2008) argues that the robo-diva character subverts stereotypical notions of both femininity and ethnicity by coming across as overtly "constructed" by technology—it thus represents a type of antithesis to naturalized conceptions of gender and/or race. In a discussion of Rihanna's video for the song "Umbrella," the lead vocal of which is opaquely autotuned, James states that by visually and sonically adopting a "womandroid" character, Rihanna renders those aspects of her appearance that usually signify her gender and ethnicity—her body and her voice—always already nonhuman. In doing so, she clearly signals that she is in control of her "self" and exploits this image in her play with what James refers to as the "neocolonial role" that society tries to attach to her as a "third-world woman ... of color, that is, as either 'Hu,' the 'human element,' or as technology/sexuality-out-of-control" (James 2008, 418). The mechanistic effect of the opaquely autotuned lead vocal clearly contributes to establishing this robo-diva subjectivity, which, according to James, is situated at "the intersection of white patriarchy's anxieties about both black female sexuality and technology" (ibid.). The subjectivity's oppositional potential lies in how each of these things is seen as potentially beyond the control of, respectively, men and humanity in general.

In all of the cases discussed above, Auto-Tune represents a dehumanizing of the human and a blurring of the traditional opposition between human and machine. In the following analysis, we will complicate this topic using two very different songs, Bon Iver's "Woods" and Lady Gaga's "Starstruck."

Hypernature and Hypernatural

Bon Iver's "Woods," from the EP *Blood Bank* (2009), is characterized by a peculiar lyrical atmosphere that is closely linked to the use of a clean, opaquely autotuned vocal that soon replicates itself into a digital choir. This a cappella performance is organized around a melody consisting of four phrases ("I'm up in the woods / I'm down on my mind / I'm building a still / to slow down the time"). The melody is repeated twice in each verse/chorus; in the following, we will refer to two such repetitions of the melody as one "round."

This goes on for five and a half rounds, meaning that the same melody is repeated eleven times. The first round (or two repetitions of the melody) is monophonic, and the visual representation of the soundwave and spectrogram reveals heavy use of digital pitch correction (see figure 7.1). Measurements of the exact distances between the different phrases in each repetition of the melody indicate that the first repetition of the melody has been looped and used as a point of departure for all successive rounds, because the timing of each repetition is precisely the same.

In each new repetition, however, new voices performing harmonies are added. In the second round the lead vocal is accompanied by one voice; in the third, by two; in the fourth, by three; and so on. The first voice sings the first phrase of the melody, f#–f#–g#–a#–e#, ending on the seventh of the F# major scale. The first harmony voice starts out on the note a# and duplicates the main melody an interval of a third above it. The other voices

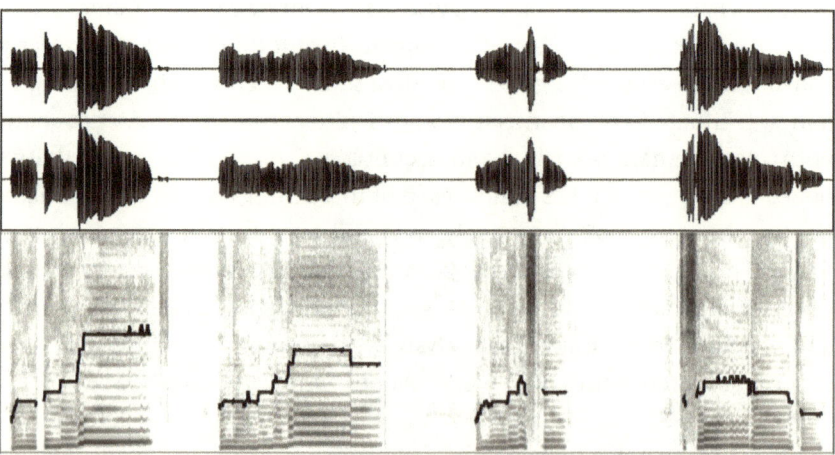

Figure 7.1
Waveform (amplitude/time) and spectrogram (0–7,000 Hz) of the first repetition of the basic melody of Bon Iver's "Woods" (Praat, version 5.1.03). Pitch contour marked by bold line in spectrogram.

mainly add more notes both below and above these two main voices, forming a chord progression that could be transcribed as F#maj7–F#add13–C#7–D#m. Because a new voice appears upon the commencement of each of the eleven rounds, the texture at the end of the song is rather dense: the sound box has been filled to its limits, or perhaps even a little beyond, causing a whisper of distortion during some of the loudest sounds.

Toward the end, a last few improvised, at times almost hysterical-sounding vocal parts appear in the higher registers. Some of these parts are colored with melismas, which, given the heavy use of digital pitch correction, jump from note to note in a "square" fashion and thus come forward as rather strange (see figure 7.2).[7]

There are several ways to interpret the sound of the digital choir in "Woods." Given the well-publicized fact that Bon Iver's debut album *For Emma Forever Ago* (Rogue Records, 2007) was recorded in a remote cabin in the woods, the content of the EP *Blood Bank* has also been associated with his experience of being isolated from other people and separated from civilization. Kanye West certainly capitalized on this association for his song "Lost in the World," which relies on a sample from "Woods," and a comment about Kanye's song on the fan website rap.genius.com, for example, regards "up in the woods" as an obvious reference to "how

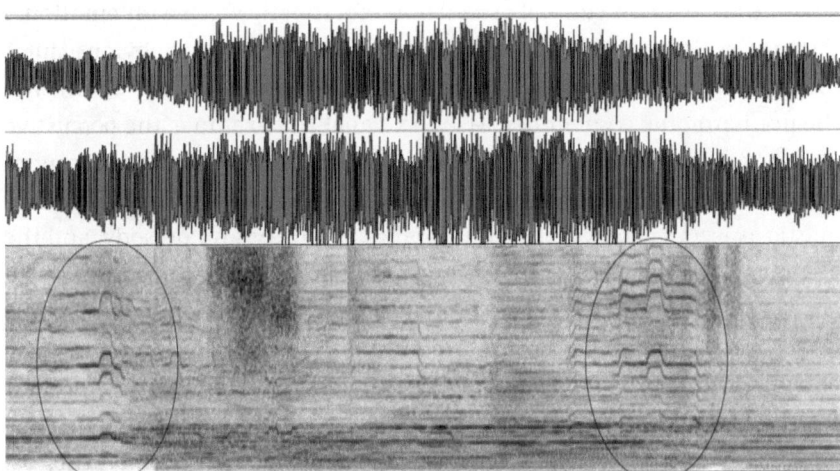

Figure 7.2
Waveform (amplitude/time) and spectrogram (0–7,000 Hz) of fragment (3:44–3:49) of Bon Iver's "Woods." Digital melismas indicated by circles.

singer-songwriter Justin Vernon ... spent all of Winter in 07, locked up in a remote cabin in Wisconsin following the breakup of his old band and his girlfriend" (comment on "Lost in the World" at http://rap.genius.com/Kanye-west-lost-in-the-world-lyrics#note-64408 [accessed September 30, 2014]). The fan then links Vernon's experience with Kanye West's self-imposed exile while recording *My Beautiful Dark Twisted Fantasy* (Roc-A-Fella Records, 2010). We might then hear "Woods" as a portrayal of a troubled individual undergoing a self-imposed isolation, and the digital choir might be the protagonist's inner voices, multiplying and growing stronger as he grows more desperate. Music journalist Ray Cummings (2010) offers a related interpretation, describing the transformation from "Woods" to "Lost in the World" as if "a *Walden*-esque shit-fit ('Woods') metamorphosed into a (probable) disquisition on man's alienation from his fellow man, how terrible we as humans are at understanding and accepting and sympathizing with one another ('World')—'lost in the woods' giving way to 'lost in the world,' a minimalist expression of angst giving way to a maximalist expression of anxiety so totemic and accessible that it threatens to overshadow everything else Bon Iver *and* West have achieved heretofore" (Cummings 2010).[8]

Here again, Bon Iver's "Woods" and Kanye West's sampling of it, in tandem with T-Pain's and West's own experiments with Auto-Tune, demonstrates the clear pop-cultural association of a robotic voice with emotional distance or flatness. The physical resemblance, of course, is obvious: auto-tuning reduces the variation in sound at the microlevel and also diminishes the nonharmonic content of the sound signal (for example, the percussive or consonant sounds of vocal utterances) in favor of its harmonic content (the periodic frequencies or pitched sounds). The neutralization of paralinguistic aspects, such as variation in timbre, intensity, and prosody, and the absence of expressive markers such as grunts, sighs, breathing, and so on are likely to be heard not only as a dehumanizing of the voice (see also the discussion of human versus nonhuman sound in chapter 3) but also as an absence in vitality and "liveness." Consonants are particularly important for communicating text as part of a vocal performance, and their presence thus connotes an interest in communicating/engagement, whereas, for example, a vocal performance consisting mainly of inarticulate vowels might be received as more aloof or introverted. Moreover, consonants and other expressive vocal sounds also convey a feeling of intimacy, because such sounds are rich in transients and die away quickly.[9] All of these aspects suggest why we hear the autotuned voice as the sound of a sad machine. In the context of loss or trauma, it is almost as though the machine becomes emotional or even empathic.

This perceived sadness might well represent an important aspect of Bon Iver's "Woods," given the line in the lyrics that states, "I'm down on my mind." At the same time, however, there is something utterly hypernatural about Justin Vernon's voice as well, not only because it is autotuned but because it is replicated into a fully digital choir. Rather than Justin Vernon himself, we hear his robotic, alienated alter ego expressing feelings and claiming a type of "first-person authenticity" (Moore 2002). This opens up, in turn, for an alternative and perhaps more optimistic reading that finds some support in Justin Vernon's own statements about the content of the album as a whole. Whereas the debut album *For Emma Forever Ago* deals with cold and darkness, loss and trauma, the subsequent EP *Blood Bank* is, according to Vernon, about the warmth that is needed to survive those conditions. In the liner notes he describes *Blood Bank* as an interpretation of shifting seasons: "These four songs explore the darker and lighter natures of the seasons and what they signify" (Bon Iver 2014). Following this "hermeneutic clue," one could frame "Woods" as a rendering of some aspect or version of nature. The cleanness of the digital choir might, for example, be thought to portray the sharp light of a white, crisp, and cold winter landscape (alluding also to the artist name Bon Iver, which is phonetic notation for "good winter" in French). However, the digital choir also evokes a feeling of distance and hyperreality, in that there is a total absence of the impure, chaotic, and disturbing aspects of real nature (in this case, the unmediated human voice). The autotuned calmness of Vernon's multiplied voice also contributes to evoking this sense of nature as perfection—that is to say, we hear nature *as culture*, or nature as a means of getting in touch with one's authentic self. Going further down this path of interpretation, there is also an obvious association here with American romantic writer Henry David Thoreau's book *Walden; or, Life in the Woods* from 1854 (which is mentioned in Cummings's article above, as well as many other reviews of, in particular, Bon Iver's *Emma Forever Ago*). Much like to Bon Iver's self-imposed cathartic time of isolation in the woods, Thoreau spent two years living simply in a cabin in natural surroundings. *Walden* is a description of and reflection on his experiences and is clearly a variation on the romantic theme of coming closer to the essential aspects of life by leaving the "inauthentic," superfluous lifestyle of civilization behind.

Interestingly, in this context, the digitally pitch-corrected voice, whether used to portray emotional distance or to introduce the possibility of a perfected nature as a means of overcoming that distance, is heard as very expressive—so much so, in fact, that it is better at evoking this spectrum of human feelings than the human voice itself. The morphing of the robotic with the human presents a humanized machine that can be used to express conditions of alienation, numbness, emotional distance, or flatness, but it does something else as well: it takes on the role of the heartbroken or downtrodden protagonist's comforting alter ego.

According to Nicola Dibben, when we emphasize the vulnerable aspects of technology, we humanize the machine. In her analysis of the meeting of nature and technology in Björk's music, she claims that Björk, by "focusing on technology's 'magical' character … undermines the idea of technology as systematic and rational" (Dibben 2009, 85). She finds that Björk's use of technological glitches is key to her subversion of many of the ideas traditionally associated with digital technology, because it is "the very unpredictability of … mechanical sources that humanizes them" (ibid., 83). By letting the technology fail and betray a lack of control, Björk "presents a more benign view of the technological, in which human and machine (organic and inorganic) are united" (ibid., 95). The poor, stiff phrasing of the human–machine singing in Bon Iver's "Woods"—and not least its robotic melismas toward the end—recalls this subversive humanization as well. Melismas traditionally characterize very expressive African American vocal traditions like soul and gospel and evoke strong, emotional vocal performances. However, when the human–machine tries to sing melismas, it sounds sort of helpless, because an autotuned voice utterly lacks the flexibility required to perform such ornaments. The "square," mechanistic melismas of the digital choir in "Woods" come forward as halting attempts to mimic a highly skilled human voice in a loaded, emotional context. Through this digital tool, then, we can generate new forms of expressivity that are typical of the "machine"—forms that shed new light on the human voice's expression of those same human feelings.

A similar emphasis on the surface and the manufactured is present in Lady Gaga's robo-diva posture in "Starstruck (feat. Space Cowboy and Flo Rida)." The song starts with a heavily autotuned vocal uttering fragments and words referring to the record production process and the star persona of Lady Gaga ("Groove. Slam. Work it back. Filter that. Baby bump that track ... Gaga in the room. So Starstruck, Cherry Cherry Cherry"). On a general level, the song comments on the staging of Gaga the star, and the video in particular overtly exposes how Gaga consciously exploits and comments on the role of stereotypical femininity in the pop circus, exaggerating it to the extent that it tilts over to the monstrous. In the words of Stan Hawkins, "Lady Gaga's construction of femininity threatens by exaggerating stereotypes, which, arguably, becomes a strategy for embodying the maternal and nurturing her monsters" (Hawkins 2014, 16). Accordingly, the soft-porn poses of Gaga the pop star as she performs the role of a diva staged for "the male gaze" are framed by a dark, evil character that apparently shows what the pop star "really" is. This "behind-the-curtains Gaga" is obviously just another stereotype, but the relationship between the two within the fictional world of the song and video signals that it is the star persona that is a pose.

The "evil" Gaga is dressed in black, wearing glasses and starts off the video with the words: "Pop—music—will—never—be—low—brow." The digitally pitch-corrected, robotic voice delivering the lyrics ("Groove. Slam ...," see figure 7.3) emphasizes further the cunningness of evil Gaga that informs pop-star Gaga: she is portrayed as an insensitive, devious machine, a robo-diva in total control of the situation who spins the supposedly naïve and male spectator around her finger with stupid poses and stunning femininity, all the while fully aware of the effect she has.

Using Simon Frith's (1996) theorizing of personas in popular music, we could sum up the layers of staging in "Starstruck" as follows: the real person (Stefani Joanne Angelina Germanotta) who works under the stage name Lady Gaga (artist persona) plays a fictional character (the song persona—that is, the evil character in "Starstruck") using her role as a pop star (the song persona's alter ego) to vague but indisputably dark ends. In his discussion of "cyborg singers," Nick Prior (2009) points to a related use of Auto-Tune on Britney Spears's "Piece of Me" (*Blackout*, Jive/Zomba, 2007). Here again the autotuned voice is used to stress the distance between the real person and the artistic persona. In contrast to Lady Gaga's playful approach, however, this example is by Prior described as "a media confessional" in which Britney self-stages her persona as a commodified "thing"—"cut up, twisted and sold back as spectacle" (Prior 2009, 12). In contrast, Lady Gaga uses the

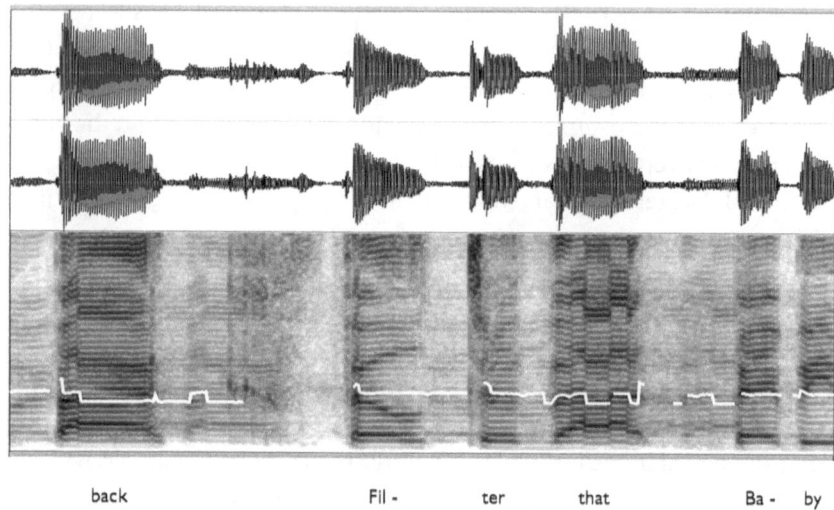

Figure 7.3
Waveform (amplitude/time) and spectrogram (0–7,000 Hz) of fragment (0:02–0:04) of the autotuned voice of the opening line of Lady Gaga's "Starstruck." Pitch contour marked by white line in spectrogram.

various layers of her artist persona to tease the spectator, forming different roles that are at the same time inseparable, as Lady Gaga constantly merges, blurs, and transforms them during her performance. According to Hawkins (2014), this is exactly what audiences have come to expect from her.

One important aspect of Lady Gaga's "Starstruck" is that it subverts the still prominent and treasured belief—at least in many popular music genres—in the existence of a certain relationship between the life and the work of the artist. It does so by staging this relationship in the song, foregrounding this possible link between the singer's star persona and the real person behind it in the fictional narrative of the Gaga video. Of course, access to the "really real" Gaga is completely blocked in this video, and the autotuned voice has an important part to play in that. Traditionally, the sound of the voice has represented the hallmark of the "real" person behind even the most obviously staged pop acts. The sound of one's singing voice is unique and, especially if it is perceived as unmediated, it is regarded as an expression of the real feelings or the real personality of its "host." Timbre and other expressive aspects of the voice have an important function in establishing this connection. The voice may thus be heard as a (perhaps unintended) revelation of the person behind it. When the voice is manipulated and takes on a mechanical or robotic sound, this link to the real "self" of the singer is broken. This might in turn explain why manipulating the timbre of the voice is so controversial and has such a radical effect. The almost hateful tone of the discussions about Auto-Tune on the Internet is probably partly due to the critical role of the unmediated (that is, transparently mediated) voice in popular music sound.

According to J. Jack Halberstam's book on what he has termed "Gaga feminism," Lady Gaga's artistic project is "a monstrous outgrowth of the unstable category of 'woman' in feminist theory, a celebration of the joining of femininity to artifice, and a refusal of the mushy sentimentalism that has been siphoned into the category of womanhood" (Halberstam 2012, xiii). Another Gaga scholar, Juliet Williams, agrees, noting that Gaga "troubles the very possibility of distinguishing original from copy, essence from performance, self from expression" (Williams 2014, 33). As demonstrated above, "Starstruck" can be framed in exactly this way: Lady Gaga clearly uses denaturalization as a central strategy when commenting on her pop star persona. Moreover, her use of parody in portraying both the pop star as she appears when staged for the male gaze and the evil monstrous personality that allegedly "hides" behind that flashy surface has an effect that is very similar to what Judith Butler claims to be the result of "gender parody": it reveals that the original identity after which it is fashioned is itself an imitation without an origin.[10]

This emphasis on the surface and the manufactured becomes even more provocative when Lady Gaga's obvious staging of her voice encounters a culture in which concerning oneself with the "surface" is inherently suspicious, and where the voice plays a particular role in revealing that which the surface is thought to conceal. The heat in the disputes regarding Auto-Tune is clearly linked to this fact: it is not a musical instrument that is mechanized here but the voice itself—that is, the expression of the human per se, or of the "real" self or human "soul." This understanding of the voice as expressing something other and more important than itself presupposes some kind of separation within the subject and, along with that, a metaphysics of inside and outside.[11] Following Jacques Derrida, the voice—or as he puts it, the voice that one hears upon retreating into oneself "in the intimacy of self-presence"—holds a privileged position in this metaphysics of inside and outside (Derrida 1973). Because one's inner self should be unified, we also expect coherence in the total expressive outcome of an artist—a coherence that somewhat tautologically is taken to refer to such a unified "real" self.

When the vocal performance of an artist is staged as it is in Lady Gaga's "Starstruck," we have no voice that can supply a reference point against which the "exterior" Gagas that we encounter in her music can be measured.[12] By way of this endless regression of staged personas, she explicitly withdraws from the authentic expression generally implied by (and expected of) the pop singer. Such a denial of the whole construction of inside and outside in popular vocal performance is, again, particularly provocative in the context of the general distrust of exteriority that characterizes the dominant tradition of Western thought. In music, as we have seen, this distrust materializes as a resistance to the technical-rhetorical aspects of the production of musical meaning (here understood to encompass nonverbal as well as verbal musical iterations). The obviously staged voice reminds us of the ways in which mediation always already colors musical meaning, or the "pure" voice of the artist (Derrida 1974, 17). When the voice is understood and valued as an expression of the "inner" self, the mediation should be *exterior* to what it mediates. According to Derrida, however, mediating tools, whether technological or rhetorical, are perceived as bad precisely because they do *not* leave that which they mediate, the pure voice or true meaning, alone. And it is that disruption that Lady Gaga uses to such pointed effect in her song, both visually and aurally through the application of Auto-Tune. A related perspective on Gaga's withdrawal from authentic expression is that she claims what Lawrence Grossberg (1992) has theorized as authentic inauthenticity. In the spirit of David Bowie and other art pop-inspired pop artists, she reveals the truth behind "true" expression—namely that there is no trustworthy, authentic, or natural position from which to judge her artistic persona(s).

Also important here are the ways in which technology in general can be used to disturb the poles of the binary opposition of nature and culture, for example to transgress stereotypical notions of race and gender. This was a big part of Prince's turn toward an avant-garde, technology-based sound with his song "Kiss" in the mid-1980s (see chapter 3). The radical use of technology was crucial to this move because it disrupted a reading of his de facto reactualization of his background in black dance music as simply a turn toward "authentic" black rhythms. Rather than being associated with "authentic" black music,[13] "Kiss" came forward as a color-free and imperatively modern manifestation of funk. In fact, neither Prince's "Kiss," Lady Gaga's "Starstruck," nor Rihanna's "Umbrella" lends much support to the usual gender and race mapping of culture/man versus nature/woman, or culture/white versus nature/black. The black artist in the cases of Prince and Rihanna and the female artist in the cases of Lady Gaga and Rihanna emerge as positioned at the pole *opposite* nature, as a cultural artifact and a product of technology. Thus these artists may all be said to explore technology in marginalizing those aspects of their performances that most effectively signify their "nature" as black and/or female artists. In the case of Prince, it was black rhythms and a raw, unmediated sound that had to be avoided.[14] In the case of the robo-diva, it is a naturalistic, "real" fabrication of the body and the voice.

The Machinized Human and the Humanized Machine

The music we've discussed here by Lady Gaga, as well as Rihanna and Prince, shows how Auto-Tune's connotations of the robotic and nonhuman can be exploited to produce a lack of depth that consciously or unconsciously furthers an aesthetic strategy dedicated to escaping unwanted essentialist readings. The alternative interpretations of Bon Iver's "Woods," presented above, demonstrate, on the other hand, a related but nevertheless distinct application of the robotic and non-human character of Auto-Tune, namely to denote alienation. When used in this context, the autotuned voice takes on the sound of a sad machine, thanks to the common association of monotone sounds with fatigue or depression. On top of this, there is something utterly surreal about the multiplication of Justin Vernon's voice into a digital choir. Rather than depicting real nature, it suggests a type of hypernature, and in this a potential cure for alienation or depression—it evokes *the idea* of nature as a means of getting in touch with one's authentic self.

Just as the microrhythmic manipulation enabled by digital audio workstations has produced new gestural expressions, Auto-Tune has complemented the human repertoire with new sounds. Digital pitch correction, then, represents yet another important development in the merging of human and machine that digital technology has brought about in the field of music. It is indeed fascinating that this technology—which is so inhuman and mechanistic in so many ways—can capture certain human states or conditions better than the unmediated voice, the most human of all instruments. When confronted with these musical artifacts, the old division between nature and technology appears to be on the verge of collapse.

In this chapter we have discussed the ways in which Auto-Tune represents a new and radical stage in the interaction of human and machine in the digital era of popular music history—a stage that has seen a decisive undermining of the traditional separation between human and machine in music production. We have also focused on some of the sonic effects that this technology can produce and how they might be interpreted. Auto-Tune indeed demonstrates how a digital signal-processing effect intended for perfecting music has become a very powerful, creative tool through literally morphing the sounds of human and machine. In contrast to the hybrid sound of the Vocoder, which presents itself as a speaking synthesizer, an autotuned voice is still a voice, albeit one that has been dehumanized. It is probably precisely the manner in which Auto-Tune makes the "human" take on the features of a machine that feels so appropriate for the expression of "the isolation, paranoia and longing of 21st-century city life" (Richards 2008)—for Kanye West and Bon Iver alike.

The fact that Auto-Tune has survived despite all the controversy it has generated has to be ascribed to both its ability to perfect the pitch of the human voice and to the ways in which it has added genuinely new sounds to the human musical repertoire. We have seen how these sounds can be used aesthetically to block the access to the "real" self of the singer, which, in turn, allows the singer to transgress traditional notions of and distinctions between man and woman, black and white, culture and nature, and to express human states such as emotional distance, numbness, and the absence of presence in one's life in a new way. As a consequence of these abilities, Auto-Tune has already become no less than a trademark or signature sound of contemporary pop music, particularly over the last decade. Even the author of the article bearing the loaded title "Seduced by 'Perfect' Pitch: How Auto-Tune Conquered Pop Music," Lessley Anderson, admits, "The glitchy Auto-Tune mode seems destined to be remembered as the 'sound' of the 2000s, the way the gated snare (that dense, big, reverb-y drum sound on, say, Phil Collins songs) is now remembered as the sound of the '80s" (Anderson 2013). Auto-Tune is here to stay, but the jury is still out as to whether it remains the ugly duckling or the beautiful swan of digital signal processing.

8 Popular Music in the Digital Era

The distinctive signatures of digital mediation have contributed significantly to the aesthetics of popular music. This book, which, of course, could not be exhaustive in this regard, has focused on some seminal digital tools that have made their mark on the popular music of the last three decades or so—the digital signal-processing effects of reverb and delay, MIDI and digital sampling, the characteristic "digital" silence, the cut-and-paste tool of digital sequencer programs, the digital "glitches" that sometimes accompany use of this virtual tool, uniquely digital forms of microrhythmic manipulation in the digital audio workstation, and autotuning. These signatures were explored via in-depth analyses of music by Kate Bush, Prince, Portishead, Los Sampler's (Uwe Schmidt), Squarepusher, Snoop Dogg, Bon Iver, and Lady Gaga, all of whom capitalize on the aesthetic potential of these signatures in interesting and compelling ways.

In the present chapter, we will revisit the analytical chapters specifically in terms of the three strands of musical meaning that we singled out in the introduction: (1) the ways in which the digitization of technology offered a new compositional palette; (2) the ways in which the digital era has generated a renewed sense of space and time; and (3) the ways in which digital technology has once again brought the issue of human versus machine to the forefront. We will then discuss the experiential tension that often arises in this context between our ecologically and historically informed ways of listening (which are constantly challenged by new musical forms) and the processes of naturalization that constantly act to change those ways (making those new musical forms historical, for example). Finally, we will speak to the significance of technological mediation in popular music, and the impact of digitization on popular music sound.

A New Compositional Palette for Popular Music Making

In our analyses, we engaged in particular with the question of why certain sounds and techniques are often conceptualized as signatures of digital mediation, how their aesthetic potential has been explored in the making of popular music, and how digital production processes have contributed to the experiential meaning of the music in our particular examples. In chapter 2, we looked at the ways in which digital delay and reverb increased the music maker's options for fabricating musical spatiality. Digital delay, as opposed to analog delay, does not automatically deteriorate in sound quality or reduce in volume, and digital reverb is completely "clean," in contrast to, for example, the metallic sound of the analog plate reverb (moreover, its reflections can be made more distinct than those naturally produced by actual spaces). These digital processing effects also gave the producer more control than ever before in terms of being able to alter any aspect of the sound with the turn of a knob or the push of a button, up to and including various manufactured presets. This general enhancement of performance and ease of design and use have allowed music makers to use these effects to mimic worldly spatial environments better than ever before. Crucially, they have also allowed music makers to *reinvent spatiality* altogether.

We exemplified the uniqueness of digital delay and reverb through an analysis of Kate Bush's "Get Out of My House," which is characterized by an almost unhinged exploitation of these processing effects. We were particularly interested in the fact that Bush and her coproducers deliberately used these effects to propose a spatiality premised not on the "real" world but on exclusively technological motivations including the delayed sonic clones of the guitar, the gated reverb of the drums, the reversed reverb of the vocal, and the spatial collage of the many vocals and instrumental sounds in juxtaposition. While some of Bush's otherworldly effects could have been produced with analog technology, they achieve unprecedented prominence here thanks to their digital origins and context, to the extent that they become, in short, signifiers of the digital. We also looked at the ways in which the track's variety of digital spatialities might affect the listener's interpretation of the musical meaning of "Get Out of My House." Its many spatial environments both support the meanings already communicated by other musical aspects of the track and generate new meanings of their own.

In chapter 3, we first discussed the ways in which digital synthesizers and samplers offered a new quantitative richness *of sounds*, as well as a new qualitative richness *in sound*. Compared to those sounds produced with analog equipment, the synthesized and sample-generated sounds of the new digital tools were significantly more complex and also represented a new presence and clarity in the upper frequencies as a consequence of the absence of noise and distortion. In combination with the MIDI protocol, which made it possible for digital music instruments and sequencers to be connected to and controlled by each other, these developments resulted in an unprecedented hyperaccuracy in the temporal domain that governed the whole groove, not only the drum parts. This temporal accuracy, together with the new sonic clarity or "realism" in the sound, would define the digital groove of the 1980s.

Then we turned to how Prince and David Z exploited these new opportunities in "Kiss," a song that demonstrates the pleasures of a machine that grooves—that is, a machine that creates compelling new dynamics by affecting both the sound and the timing of the music at the microlevel. Prince and David Z managed to get the machine to groove in this way by tweaking its capacities to their limits. By using the hi-hat track as a trigger for the guitar, they produced a guitar track that no longer sounded like a guitar but instead presented chords and chord changes of uncertain instrumental origin, pulsing along with the hi-hat pattern. Though this leads to an unmistakable "programmed" feel, it is combined with a very dynamic use of sound in a manner that was truly remarkable at the time it was produced.

In chapter 4, we first framed the characteristic silence of digital technology in terms of its radical departure from its predecessors' various noise by-products. By fulfilling the cultural and historical quest for complete transparency or high fidelity to the sound source, it provided a new basis for comparison with older media that, rather perversely, inspired many people to embrace them because of (rather than in spite of) their noise. Not only did this type of noise become more "present" in the age of its absence, then, but it also became revitalized and revalued. What were once regarded as limitations of technology were now redeemed as desirable aesthetic qualities and, in turn, new compositional opportunities. These opportunities—which involved the exploration of predigital medium signatures through old equipment, obsolete recording techniques, or samples of predigital music—were embraced most enthusiastically in the 1990s, which in effect became a decade characterized by a lo-fi movement that crossed genres and styles.

We found, however, that instead of rejecting digital technology altogether, music makers who subscribed to this trend often tended toward various combinations and interminglings of old *and* new technology. Their music, while understood to be speaking to (and speaking in the voice of) the past, was in fact firmly situated in the present. A case in point is Portishead's "Strangers," which our analysis revealed to be characterized by its juxtaposition of sonic signatures of both previous and present musical eras. For example, the medium noises in "Strangers" are disrupted by moments of utter digital silence, and this jarring contrast makes us encounter both the analog and the digital afresh.

In chapter 5, we discussed how the virtual cut-and-paste tool of digital sequencer programs renewed and reinvigorated the experimental cut-and-paste approach by eliminating the extremely time-consuming process of physically splicing analog sound tapes. Moreover, the level of precision in digital cut-up operations—thanks to the ability to zoom in on the sounds and treat them at a microlevel—is unprecedented. Like digital spatiality, MIDI and sampling, and digital silence, then, digital cut-ups augment the compositional palette of music makers, in terms of new sounds and new musical effects, and in so doing also augment the aesthetic preferences of music listeners.

Through our analyses of "La Vida es Ilena de Cables" by Los Sampler's and Squarepusher's "My Red Hot Car," we examined in detail how this form of technological mediation contributed to the sound and groove of the music, and ultimately to its meaning and aesthetic appeal. In our discussion, we were particularly interested in how these sounds are reminiscent of digital malfunctions such as the stuttering sounds caused by a CD player that cannot read the information on a scratched disc or the hiccups and pauses caused by buffer underruns during a computer program's playback of audio files. Because cut-ups and other glitch sounds are now rather common, they have lost most of their shock value, but, thanks to their profound associations with technological failure, they continue to draw attention to themselves as acts of opaque mediation. This consequently allows for their double experiential meaning: they both are and are not "part" of the music, and are at once desirable and undesirable as a consequence.

We also discussed how "La Vida es Ilena de Cables" and "My Red Hot Car" might be conceptualized as music within music, as if a "normal" or traditional layer were being interrupted or manipulated by a layer of cut-ups and glitches. Each layer's meaning and function relies on the other: we would not experience these sounds as cut-ups, medium glitches, or signal dropouts if we were not aware that there was something to be glitched or missed; conversely, the glitches reinforce (or at least evoke) our association of "music" as such with a spatiotemporally coherent singular performance. Again we meet with the double meaning of this music: these tracks both are and are not traditional performances; their glitches both disrupt and constitute the music. Cut-ups thus draw attention to both the recording/production medium's ability to transparently mediate a message *and* the medium's constant participation in that message.

Digital tools have also enabled the profound manipulation of rhythm, not least at the microlevel. In chapter 6, we approached those "seasick" grooves of contemporary R&B and hip hop that became a fad among producers at the turn of the millennium. Common to these grooves is the fact that their "feel" aspect is almost overdone. The trend evoked the "deep" funk grooves of 1970s funk bands, though this onetime organic feel was given a distinctive computerized update through digital manipulations and in particular the new possibilities for microtemporal editing brought about by digital recording. The temporal discrepancies between the different rhythmic layers forming the repeated pattern of these seasick grooves simply transgress the perceptual capacities of a musician, leading to the impression of an exaggerated rhythmic expressivity. These new feels were made possible by the digital audio workstation, which provided extraordinary opportunities for controlling and manipulating the temporal aspect of music—for example, entire rhythmic patterns and/or single rhythmic events could be moved earlier or later in time with unprecedented specificity, or a sample could be integrated into the main groove and then twisted or bended, producing peculiar rhythmic effects as a result.

We approached this trend through analyses of two songs from Snoop Dogg's *Rhythm & Gangsta* album where some of these new opportunities for manipulating the rhythm of the music have made an audible mark on the sounding result. The bass riff of "Can I Get a Flicc Witchu" is an example of a rather extreme temporal/rhythmic mismatch between the main groove and a source sample or recording—so extreme, in fact, that it becomes the groove's signature element, which in turn clearly affects the overall impression of the song. In our second example, "Bang Out" from the same album, the effect is less extreme. Here, the peculiar machine-generated rhythm arises from the clash between two alternative subdivisions in the groove. Clashes between different metrical grids at the level of the subdivision represent a feature that appears in various groove-directed musics, both played and programmed. In a computer-based groove like "Bang Out," however, one can experiment with potentially conflicting patterns and even vary the microtiming in one or all of them without having to account for the perceptual and performative limitations of human beings. The bass and bass drum pattern in either "Bang Out" or "Can I Get a Flicc Witchu" is almost unplayable, at least repeatedly, and testifies to how far one can go with the microrhythmic manipulation that is enabled by the integration of digital sound editing and MIDI and audio sequencing in the digital audio workstation.

In chapter 7, we discussed how the creative use of the digital signal-processing tool Auto-Tune has become a valuable source of completely new sonic effects in popular music production. Despite harsh criticism and controversies, digital pitch correction has become a tool for creating new characteristic sounds, and Auto-Tune is a compelling example of the tendency to experiment with a correction technique aimed at achieving perfection within an existing recording paradigm. The opaque use of pitch-correction tools, also called the "Cher effect," flourished in the 2000s and has become associated with certain topos in popular music. In hip hop it is often used to express a feeling of alienation, sadness, and loss. We also pointed out how Auto-Tune's connotations to the robotic and nonhuman have been used as a strategy for escaping stereotypical notions about the nature of race and/or gender. This use of the Cher effect is a common theme in the reception of female artists within electropop and black R&B, where an extreme use of Auto-Tune is often combined with imagery depicting monstrous femininity and/or hyperembodiment, in the sense of a body that comes forward as being perfect or otherwise cultivated beyond the human.

We then analyzed the overdone digital pitch correction of Bon Iver's "Woods." The combination of the song's poetic atmosphere and the clean but artificial sound of the lead vocal evokes the relationship between nature and culture. Particularly toward the end, when improvised, melismatic, and at times almost hysterical vocal parts in the upper register are subject to heavy pitch correction, the result is a peculiar and very characteristic signature that comes forward as a weird, "mechanical" form of expressivity. Ultimately, we point to how Lady Gaga, in line with previous robo-divas' use of this effect, harnesses digital pitch correction to establish a monstrous alter ego in her song "Starstruck." In both of these songs, the heavy use of Auto-Tune transforms the sounding voice into a hybrid of human and machine that is reminiscent of previous analog instruments such as the vocoder, but yet different since the Auto-Tune seems to "instrumentalize" the voice itself.

With these various conclusions in mind, it is obvious that digital signatures of mediation, and the aestheticization of these signatures, have added to the music maker's compositional palette. As we emphasized in our introductory chapter, the signatures studied in this thesis are not all of those that exist, but they nevertheless serve to demonstrate that the digitization of technology has affected popular music aesthetics in important ways. Perhaps most obviously, it has introduced sounds and effects that are now giveaways to its presence, such as complete operating silence, sonic clones (as in delayed sounds), the sounds of digital glitches, machine-generated rhythmic and vocal expressivity, and new forms of sonic spatiality.

In addition to the musical changes brought about by qualities that are *unique* to the digital medium, other sonic features have resulted from the ways in which digital technology has reinvented analog tools and techniques. The digitization of sampling and cut-and-paste techniques, for example, brought about a quantitative change in their impact on popular music production; while certain musical features are not unique (such as the ability to cut and paste music), the scale with which these features are now applied is almost unthinkable via analog technology. The fact that the digital versions of these tools and compositional techniques are all based on fundamentally different technical principles has resulted in significant alterations to their functions, applications, and sonic results, as discussed above. Likewise, the fact that these once-analog features are so much more present in digitally produced music means that they must now be regarded as signatures of digital mediation in particular—at some point, that is, a quantitative change becomes a qualitative signature.

Digital mediation's qualitative and quantitative changes have affected popular music aesthetics tremendously in terms of their contributions to entirely new musical effects. This new compositional palette provided by digital technology has in turn encouraged an approach to music composition that privileges spatiotemporal experimentation and a sonic acknowledgment of the spatiotemporal disjuncture of the sounds. It is to this discussion that we now turn.

A New Era of Schizophonia

For centuries, all forms of music had particular features in common: the music was always performed by musicians located in one specific place, and it unfolded organically over time. Accordingly, music was, without exception, spatiotemporally coherent and could *only* be heard in accordance with the acoustic laws that applied to its "live" performance. These specific and defining qualities of music did not change until the invention of the phonograph in 1877, which introduced people to an era of what Canadian composer and writer R. Murray Schafer has labeled *schizophonia*, to emphasize the distinction between original and reproduced sounds (*schizo* is "split" and *phonia* is "sound" in Greek).[1] Schafer characterized schizophonia as a permanent and uniform condition, but we find it relevant to nuance it in relation to three musical eras: the mechanical, the magnetic, and the digital.[2]

The invention of the phonograph occasioned the cultural shift to schizophonia (in its mechanical era) in terms of splitting sounds from their spatial and temporal origins and thereby challenging our traditional understanding of sounds as emerging directly from a live source. However, although the sounds of a musical performance were cut loose from their origins in time and space, a lack of contemporaneous editing possibilities largely restricted the phonographic recording medium to perpetuating the notion of music as a spatiotemporally coherent performance: what you heard on the recording was the sound of a preexisting coherent event that had been recorded in a single take. The rare exceptions to this were recordings that resulted from very early applications of the technique of overdubbing. Still, the socially ascribed meaning and function of the phonograph was generally to serve as an archival medium rather than a creative tool. Thus, while the possibility of manipulating time and space in the reproduction of music had existed since the birth of the recording medium, it was only through the magnetic tape recorder and the invention of multitracking that it became truly viable.

The invention of the magnetic tape recorder brought about a new schizophonic era, thanks to its dramatic new possibilities for the spatial and temporal disjuncture between sound and its source(s): fundamentally, it allowed for several takes that could be treated separately, which in effect enabled the editing of individual aspects of a given performance.[3] For example, tape made it possible to literally cut tracks apart and paste them together again through the process of splicing, allowing engineers to juxtapose musical tracks from different times and places. The spatiotemporal disjuncture of sound was further ushered along by the advances of the multitrack recorder over the one-track recording machine. This technology made mistakes less disruptive, because the individual parts in the music were stored on separate tracks. In addition, parts could be recorded separately at different times and, if desired, in different locations, and the multitrack recorder solved the problem of degradation in sound quality that took place after each new overdub. Also, because sounds could be recorded through several channels without being automatically bounced onto a single track afterward, the tracks could be treated separately even *after* they had been recorded. Because of its new recording and editing capabilities, then, the recording device transitioned from an archival to an artistic medium, one that represents musical performances with no claim to their preexistence. Consequently, recorded music in the magnetic era came to encompass a patchwork of sounds recorded at different times and in different spaces, in turn challenging even more strongly than previous recording technologies our aural sense of time and space.

If the phonograph split sound from its source and the magnetic tape recorder split the bundle of recorded sounds from their shared spatiotemporal frame, what was left to the latest era of schizophonia? An ever-expanding economy of scale: though digital technology did not split sounds any further from their sources than the magnetic tape recorder did, it allowed the act of splitting to be both more profound and more frequent, even to the extent of inaugurating yet a third era of schizophonia. Along with its enhanced capacities for splitting and manipulating sounds, digital technology also allows the music maker to undo what is done—it is not, in and of itself, destructive to the materials on which it acts. This has in turn encouraged a more experimental approach to music making, and to manipulating the music's spatiotemporal form. Although music that exposes its spatiotemporally fragmented form has a long history, digital mediation represents a fresh take on the whole art and craft of popular music production in these terms and others.

The music analyzed in this book exemplifies different digital means of exploiting the spatiotemporal disjuncture of sounds. In chapter 2, we discussed music as a sonic montage of sounds that seem to belong to different spaces or sonic environments; here, the production represents a purposefully *spatial* schizophonia that foregrounds the act of splitting the sounds from their original spatial settings. In chapter 3, we touched upon the hyperreal aspect of the overall sound of "Kiss." The extreme clarity and presence made possible by the digital processing of sound are first of all used to create the profound proximity of Prince's voice. The hyperpresence of his voice is further underlined by the surrounding digital "nothingness": there is no noise, distortion, or filler in the sound. The dryness of the overall production also contributes to its projection of a virtual space of almost surreal dimensions, owing to the lack of depth in the sound box. In chapter 4, we discussed how digital silence encourages music makers to revisit sounds originally belonging to the past, split them from their sources, and insert them into a new musical context; the schizophonic musical result simultaneously evokes different eras, thanks to the characteristic material signatures of different mediums. In chapter 5, we discussed music that bears audible traces of being chopped up and manipulated by cut-and-paste operations, often simulating the distortion attendant upon technological malfunction as well, to create a schizophonic distortion of traditional temporal coherence. In chapter 6, we saw how the warping of sampled sounds can generate completely new forms of microrhythmic "feel," which in turn makes obvious the differences in spatiotemporal origins among the various elements in the groove.

Despite all of this experimentation and change, the spatiotemporally coherent form remains central to contemporary popular music aesthetics. That is, music makers have approached the schizophonic abilities offered by each recording/production medium in very different ways; some have sought to conceal the music's fragmented construction, while others have sought to expose it. This demonstrates that the perceived affordances of digital technology differ from consumer to consumer and from context to context, and that they will not always serve radical ends. Obviously, then, the increase in musical manifestations of spatiotemporal disjuncture cannot be explained by the advent of digital technology alone but requires a certain mindset as well—one that is perhaps more common now but by no means ubiquitous. However, the fact that digitally converted sounds can be treated very differently from analog sounds has certainly informed, and in some sense transformed, the ways we compose and produce music, the ways we listen to it, and, ultimately, the ways we conceptualize it.

Human versus Machine: Morphed into a New Millennium

Various consequences of the perceived (if fascinating) conflict between human and machine have underpinned the history of music in the twentieth century and beyond. Digital technology has renewed the debate in music making, whether one nurtures the machine for its own sake or seeks to make the machine imitate the human. More broadly, it has ushered in the distinct possibility that the relation between human and machine is less a bald opposition than a sort of continuous exchange. In the words of Nick Prior (2009): "It is not just that technology *impacts* upon music, *influences* music, *shapes* music, because this form of weak technological determinism still implies two separate domains. Music is always already suffused with technology, it is embedded within technological forms and forces; it is *in* and *of* technology" (Prior 2009, 95). As Prior and also Kvifte (1989) have pointed out, this is not a new situation. Playing a traditional instrument also means being deeply involved in music technology. The development addressed in this book thus might be more accurately described as follows: While played and machine-generated music started out as utterly distinct aesthetic fields, they have ended up as inseparable domains, deeply embedded in one another. Digital technology has contributed tremendously to the ongoing transformation of popular music from an "either/or" proposition to a "both/and" hybridization that makes it increasingly difficult for listeners to distinguish between human and machine-made musical utterances.

At the outset of this story—that is, prior to MIDI and digital sampling—machine music sounded quite literally mechanistic both sonically (sounds were less complex) and rhythmically (events fell regularly on a precise metrical grid). As discussed in chapter 3, there was a close association between isochronous timing and the use of machines to create music, because all of the rhythmic events were positioned on the grid of the sequencer and therefore lacked both the deliberate and the unintended variations produced by musicians. The absence of the small variations in intensity and timbre that are always present in the sound of played music also contributed to this "stiff," mechanistic feel.

An early step in the process of getting the machine to sound like a human groove performance involved the use of natural-sounding drums in the digital drum machine. As discussed in the analysis of Prince's "Kiss," the increase in sonic richness due to the use of sampled actual drum sounds became very important to the sound of this tune. An equally important achievement was the introduction of small variations in the sounds of the hi-hat strokes through the manipulation of digital delay and the MIDI-enabled connections among different devices. The resulting distinctive hyperaccuracy in timing evoked the machine but in the new context of an unprecedented richness and clarity in sound.

With the digital audio workstation, a new era in the machine's ability to affect the groove at a microlevel began. Around the turn of the millennium, some years after the software developed primarily for sequencing and the hi-fi world of digital recording had begun to converge, the potential for manipulating the microrhythmic design of recorded musical events began to be realized in African-American-derived groove-based music. It is no surprise that this tradition was pioneering in this respect. Hip hop was already tightly coupled to developments within music technology, such as the sampler and the sequencer, and the microrhythmic aesthetics of contemporary R&B, neo-soul, and hip hop also encourages experimentation with new and more extreme rhythmic feels. What was new with digital audio sequencing and editing was that relatively significant timing discrepancies between rhythmic layers could become part of the groove itself. In contrast to the Prince example, where the machine starts to groove because of a very dynamic use of sound, the machine here starts to groove in peculiar ways because the timing of entire tracks as well as single events has been manipulated using visual editing. This new ability, combined with the capacity to "undo" anything that had been done, allowed for extreme experimentation and a quantitative increase in the capacity to complicate rhythms—an increase, again, that eventually produced a qualitative change in the music. As was demonstrated in the analysis of the songs from Snoop Dogg's *Rhythm & Gangsta* album, such machine grooves are characterized by a gestural sonic repertoire that was entirely their own and never before heard as such by any human being. As discussed in chapter 6, Snoop Dogg picks up the seasick feel produced by the warped sample in "Can I Get a

Flicc Witchu" in his rapped lead vocal. In "Bang Out," as well, the vocal bears traces of the kinds of phrasing that often appear atop clashing subdivisions, audible as a distinct in-between shuffling of the subdivisions of the beat, as though the vocal were drifting between duple and triple subdivisions. Both of these adjusted vocal alignments are examples of how people can learn, and actually do learn, from the machine.

Digital technology also facilitated the act of taking control over the machine's malfunctions and unavoidable sounds, as demonstrated in chapters 4 and 5. For example, "La Vida es Illena de Cables" by Los Sampler's and "My Red Hot Car" by Squarepusher rely on the virtual cut-and-paste tool to assume control over "unwanted" glitch sounds (such as skips, stutters, and signal dropouts), insert them in the music with milliseconds of precision, mangle them in tasteful (or distasteful) ways, and thus transform them into musical gestures in their own right. Similarly, in "Strangers," discussed in chapter 4, Portishead uses the distorted sounds that accompany earlier sound-recording technology, which have long been dismissed as unwanted but unavoidable noise, to musical ends. Digital technology is here used both to sample or reconstruct these sounds and to contrast these sounds with its own characteristic silence. The experimental use of glitches and past medium signatures makes it obvious that while malfunctioning or outdated technology might leave us helpless and frustrated, thus introducing a gap between the machine and the human, it also can inspire us creatively and even become, in turn, a means of uniting the two.

The digital tools discussed in this book can be used either to perfect an existing musical performance or to generate something completely new—to correct errors, that is, or to utterly broaden the compositional palette. These two fields of application of new technology, which we have labeled *machine-aided perfection*, which signals a musical paradigm of transparent mediation, and *machine-generated creation*, which signals a musical paradigm of opaque mediation, respectively, always go hand in hand. Quoting Richard J. Burgess: "Technology evolves to solve problems and in doing so presents potential for previously unsuspected applications. Inventors, innovators, creative artists, and hackers ... recognize these accidental or incidental capacities. ... Even people who consider themselves ordinary often find fascinating ways to misuse and abuse technology to great effect" (Burgess 2014, 136).

Auto-Tune is perhaps the ultimate example of a tool that was intended for perfecting music but in the end did so much more, by morphing the sounds of human and machine. It has sparked much controversy, probably because it comes forward as such a powerful "dehumanizing" tool: it is capable of transforming the human voice into a robotic alter ego. In contrast to the hybrid sonic character of a vocoder, which presents itself as a speaking synthesizer, an autotuned voice is still a voice, albeit one with machinelike features. As was demonstrated in chapter 7, this can be used to portray a human that feels like a machine, for example as a consequence of the alienation, longing, or other traumas of modern life. It can also be used to transform the artistic persona *into* a machine, "blocking" the access to the artist's "inner" or "real" self in order to avoid essentialist readings of the music based in, for example, gender and race, and hinder interpretations focused on that which might be thought to reside underneath the "artistic" surface.

As Burgess describes above, digital reverb and delay can also be used either to polish and perfect the acoustic qualities of a musical performance or to create acoustic qualities that sonically diverge from real-world acoustic environments, as demonstrated by Kate Bush's "Get Out of My House," discussed in chapter 2. On the one hand, we might experience these otherworldly sonic environments as completely natural, because machine-aided sounds are part of our daily lives. On the other hand, we might experience them as unnatural or surreal, as we continue to compare our sonic environments with our sonic experience with real-world acoustic environments. The distinction between the machine-generated world and the human world is neither totally absent nor totally discernible.

Digital technology has helped to humanize the machine and encouraged humans to imitate (and merge with) the machine. For example, the timing of musicians is warped in the digital audio workstation, then copied by other musicians, who are in turn manipulated in new machine-generated renderings, and on it goes. As a consequence, the rhythmic expressions of humans and machines are today, at least in some genres, so deeply mingled that it is impossible to say where one ends and the other begins. As this example demonstrates, it has, in fact, sometimes become difficult to distinguish between human and machine. Following Heidegger (1977), we could also say that technology once again has demonstrated its ability to extend human behavior. The machine does not have a life of its own, but it does spur a constant negotiation between its affordances and human creation.

Historical Listening Constraints and the Tuning of the Ear

The fact that spatiotemporally fragmented or otherwise technologically mediated music has only existed for a short while relative to the span of music's history has certain perceptual consequences: we still unconsciously frame the new music we hear using our historically and culturally deep-rooted notion of music as an unmediated, spatiotemporally coherent form. Likewise, our impression of music as technologically mediated (cut-up, glitched, or overprocessed) derives directly from the alternative: technologically unmediated music, or music that necessarily unfolds in a unified space and time. As Linda Hutcheon puts it, discontinuity is revealed at the heart of continuity (Hutcheon 1988, 11), and, we might add, mediated sounds are revealed at the heart of unmediatedness.[4] For example, we continue to understand a fragmented musical event as a coherent performance that has been disrupted, even though that performance never existed in the first place. Generally, we attribute meaning through habits or conventions, and our ways of listening are therefore informed by our historical and cultural backgrounds. While none of us lived before 1877, the traditional musical form of that time (a spatiotemporally coherent performance) remains with us today.

However, in addition to our "historical" listening constraints, another force affects our perception: the tuning of our ears. When a new sound is introduced through the musical recording as the result of a new musical tool or technique, it might initially draw considerable attention to itself, and this is because we experience it as opaque and therefore possibly as weird, surreal, supernatural, or uncanny. Yet new musical expressions often become naturalized, so that the new sound will one day be the norm against which still later sonic environments are measured. Many long-standing fabricated sounds on popular music recordings have become so naturalized that they barely evoke any sense of uncanniness whatsoever. Our elasticity as to what we regard as "natural," in other words, is enormous, thanks to the tuning of our ears. Writing in the late 1930s, Walter Benjamin already observed that what is "standard" in a technologically mediated reality may in fact be so normalized that when a mimetic representation of the unmediated reality behind it appears in this mediated reality, it may be experienced in turn as "unstandard," or even as mediated anew: "That is to say, in the [film] studio the mechanical equipment has penetrated so deeply into reality that its pure aspect freed from the foreign substance of equipment is the result of a special procedure. ... The equipment-free aspect of reality here has become the height of artifice; the sight of immediate reality has become an orchid in the land of technology" (Benjamin 1968, 233). For example, the voice in today's popular music productions is generally highly mediated, in terms of being compressed, equalized, reverbed, autotuned, and so on. This means that when we hear a voice, either on a musical recording or at a concert, that feels different from this compressed and voluminous high-definition sound, we want to blame somebody (usually the sound engineer). The unmediated voice, today, is that orchid in an otherwise utterly mediated musical environment.

In 1977, R. Murray Schafer noticed that the soundscape of the radio, whose sonic juxtaposition of different shows is indeed "unnatural" relative to technologically unmediated sonic environments, has been rendered natural thanks to the influence of other electroacoustic devices in company with it. As a result, he reasons, "the radio has actually become the birdsong of modern life, the 'natural soundscape'" (Schafer 1977, 93). Surely, then, the mediated world can be experienced as just as real or natural as any unmediated environments. Yet our awareness of alternative contexts, and of what rules apply within them, remains very strong. For example, though a technologically filtered voice may now be naturalized in a *musical* context, it would be uncanny indeed if the person next to us suddenly started speaking in that sort of voice. James J. Gibson's realization that the same environment can afford different things to people in different contexts (Gibson 1986, 128; see also chapters 1 and 2 in this book) may also shed light on how we sometimes experience the same sound as both surreal and naturalized—it depends, in other words, on the context to which we compare it. The sonic montages of spatial environments in Kate Bush's "Get Out of My House," the combination of machinelike timing and acoustic sounds in Prince's "Kiss," the evocations of different aural eras in Portishead's "Strangers," the cut-up music of Squarepusher's "My Red Hot Car" and "La Vida es Ilena de Cables" by Los Sampler's, the exaggerated expressivity in the grooves of Snoop Dogg's *Rhythm & Gangsta* album, and the morphing of human and machine in the voices of Lady Gaga and Bon Iver only come across as manipulated, surreal, or uncanny in comparison to a performance that is perceived as different (in the sense of unmediated or spatiotemporally coherent). Taken completely on their own, or in the company of other tracks from the same music maker or genre, these manipulated musical expressions might well be experienced as perfectly appropriate, and even normal and "natural" in their own ways.

The intersection of our historical listening constraints and the tuning of our ears also explains why we might experience digital sound signatures as balanced on the border between the music's interior and exterior. For example, the sequences of digital silence in Portishead's "Strangers" are simultaneously experienced as a void of sound or lack of medium signature (the music's exterior) *and* as a sound and medium signature in themselves (the music's interior). For the same reason, the noise in "Strangers" is at once understood as the unavoidable by-product of equipment from an earlier era (the music's exterior) and as an intended musical sound effect (the music's interior). Even the cut-up, glitchy sounds of the music by Squarepusher and Los Sampler's are at once received as unmusical sounds (sounds that disturb the music) and musical sounds (sounds that constitute the music). The same could be said about the extreme microrhythmic feels presented on Snoop Dogg's *Rhythm & Gangsta* album: they are at once completely out of time and perfectly in (their own) time. While the former understanding of all of these sounds derives from our historical listening constraints, the latter understanding of them derives from the tuning of our ears.

The listener's comparison of the music of a recording to both an unmediated, spatiotemporally coherent musical performance that follows strict acoustic laws *and* the contemporary, technologically mediated musical environment in which anything goes gives rise to a rather illuminating tension that yields analytical insights into the strategies behind digital music making. These two inclinations do not converge, and they do not erase one another; instead, they appear to work together in a suspended state perhaps best described as "both and neither." Ultimately, the coexistence of these two perceptual forces—the historical and cultural constraints of listening and the liberating processes of naturalization, or "the tuning of the ear"—offers compelling insight into the reasons why we find ourselves challenged, titillated, and even overwhelmed by the many striking musical forms of technological mediation of our digital age.

The Significance of Technological Mediation to Popular Music

In our introduction, we reviewed Brøvig-Hanssen's notions of transparent and opaque mediation, with which she has tried to deconstruct the binary of mediated versus unmediated. What is generally perceived to be technologically unmediated music is usually transparently (as opposed to opaquely) mediated music instead—after all, very little popular music is *completely* technologically unmediated. The nuancing of this binary also emphasizes the fact that while a given act of mediation might stay the same, people's experience of it will vary with time, place, and genre. Furthermore, the notions of transparent and opaque mediation signal alternative musical paradigms. In transparent mediation, technological mediation is used to embellish what already is; in opaque mediation, the technological mediation's self-presentation is exploited in and of itself. In the musical examples analyzed in this thesis, the technological mediation is opaque. Here, the significance of this operation is evident—the technology has a voice of its own, and insists on its part in the experiential meaning of the music. The aesthetic potential of the technological mediation's self-presentation or signature is dedicated to the production of unique musical effects, and its opaqueness is thus celebrated.

Mediating technology is actually imperative to all forms of popular music, even those that privilege transparency. And yet, as Simon Frith observes in "Art versus Technology: The Strange Case of Popular Music" (1986), certain musical contexts such as the rock genre (at least in some instances, if not in the genre as a whole) continue to regard technology as *inauthentic*, and even as a barrier between the listener and the musicians or sound sources: "The continuing core of rock ideology is that raw sounds are more authentic than cooked sounds. This is a paradoxical belief for a technologically sophisticated medium and rests on an old-fashioned model of direct communication—A plays to B and the less technology lies between them the closer they are, the more honest their relationship and the fewer the opportunities for manipulation and falsehood" (Frith 1986, 266–267). Of course, as Frith is perfectly correct to point out, this belief that technology hinders direct communication is paradoxical and, above all else, quite relative (ibid.; see also Frith 2012)—the same listener might accept certain forms of technology while rejecting others, and, as mentioned with regard to the tuning of our ears, some technological features that are initially rejected might later come to be accepted. The question, in other words, is not only how much technology "lies between" A and B but also how and to what extent the listeners perceive it.

Put simply, if we do not notice the technological mediation, it is because this mediation is (to us, at the moment) transparent, *not* absent. The genre labels "electronica" and "techno," for example, imply the application of more mediating technology than the genre label "country," but these stereotypical associations are more the result of *how* technology has been used in the genre than of *how much* has been used. When music is criticized for being inauthentic because it is too reliant upon technological manipulation, it is in fact less the mediating technology itself that is under attack than the aesthetic that privileges its opacity over its transparency. What is at stake here is the involvement of technological mediation *perceived* as mediation, not the involvement of technological mediation per se. And what is sometimes described as a lesser degree of technological mediation (and, in effect, a more authentic musical expression) should instead be recognized as transparent mediation and, correspondingly, as a rhetorical attribute or mimetic strategy that is every bit as purposeful as the alternative.

Throughout this book, we have nevertheless tried to elaborate upon the cultural currency of contemporary popular music and its technological mediation by analyzing *opaquely* mediated music. Given our overarching concern with the very aesthetics of popular music, we have tried to illuminate some of the intricacies around the ways in which technological mediations in general, and digital mediation in particular, inform and transform musical expressions (and musical experiences). In addition to demonstrating that the methodology of music analysis might be pertinent to the process of understanding the cultural significance of technological change, we also hope to have made the case for taking technological mediation into account when studying popular music.

Digital technology, as argued throughout this book, has offered relatively few operations that are entirely new. Nonetheless, the technological change from analog to digital has had a tremendous cultural impact. In addition to offering unique sounds, processing effects, and functions, the digitization of technology has made earlier practices much more straightforward. It has also changed our approach to certain predigital musical tools, to the extent that both the practices and the tools in question have been revisited and, in a sense, reinvented. While the production process of popular music continues to thrive as much on tradition as it does on innovation, the examples in this book demonstrate the profound ways in which the digitization of technology has influenced the aesthetics that underpins the art form. In fact, it is debatable whether these tracks could even have come about were it not for digital mediation and the power inherent in its revelation.

Notes

1   Introduction: Digital Technology and Popular Music Sound

1. The reality is, of course, rather more complex than this stereotypical narrative admits. For example, the new digital models of music production and consumption have not overtaken but instead come to coexist with the traditional large-scale corporate models of the industry (see Hesmondhalgh 2005, 171). Moreover, as Paul Théberge points out, it has become increasingly difficult to distinguish between the categories of the "professional"/"commercial" studio and the "personal"/"home" studio (Théberge 2012, 83).

2. Of course, as scholars such as Timothy Day and Colin Symes have pointed out, the recording medium is neither neutral nor objective in this regard; see Day 2000 and Symes 2004.

3. For discussions concerning music and copyright, see, e.g., Frith and Marshall 2004; Lessig 2004, 2008; McLeod 2005a, 2005b, 2007; Vaidhyanathan 2003.

4. See Sterne 2003 and Théberge 1997 for exemplary studies of how sound technologies are socially and economically embedded.

5. For a discussion of changing listening practices, see Bergh and DeNora 2009; Bull 2000, 2007; Ebare 2004; Krims 2010; Skånland 2011.

6. In coining these terms, Brøvig-Hanssen particularly benefited from Marin's articles "Opacity and Transparency in Pictorial Representation" (1991) and "Mimesis and Description" (2001, first published in 1988). She later discovered that her use of these terms also had much in common with their application by Jay David Bolter and Richard Grusin, who apply the concepts of transparency and opacity in a fashion reminiscent of Marin in their *Remediation: Understanding New Media* (2000). There are, however, certain differences between Brøvig-Hanssen's use of the terms and the notions of Marin and Bolter and Grusin that can be traced to the fact that while Marin, as well as Bolter and Grusin, discusses *representations* (which are based on substitutive signs), Brøvig-Hanssen discusses technological *mediation* (which is not based on signs in this sense). See Brøvig-Hanssen 2013 and forthcoming.

7. For a discussion of surreal versus naturalized effects of technological mediation, see Brøvig-Hanssen and Danielsen 2013.

8. For a discussion of the development and reception of close-up microphone singing, see Chanan 2000, 67–70, 109–110; Frith 1986, 263–265; Toynbee 2000, 74–80; Read and Welch 1977, 238–239.

9. See, e.g., Roads 2000, 7–47; Watkinson 1999, 110–122.

10. Videotape, which could handle much more information than magnetic audiotape, was often preferred for the storing of digitally converted sounds (Millard 2005, 349).

11. "Remediation," as Jay David Bolter and Richard Grusin define it, is "the representation of one medium in another" (Bolter and Grusin 2000, 45).

12. Among the leading DAW programs in the 1990s and 2000s were Digidesign's Pro Tools, Emagic's Logic Pro, and Steinberg's Cubase, and they remain viable today, though new challengers include Ableton Live, Sonar (formerly known as Cakewalk), Reaper, and PreSonus Studio One, as well as Fruity Loops and Reason (which have only recently started to offer audio recording).

13. See Brøvig-Hanssen and Danielsen 2013, where we discuss this specific aspect of digital technology in our analysis of Suede's "Filmstar" (*Coming Up*, Nude Records, 1996).

14. For a thorough description of some of the most common visual interfaces of the DAW, see Kvifte 2010, 214–219.

2 Making Sense of Digital Spatiality: Kate Bush's Eerie Collage

1. Kate Bush, quoted in Shearlaw 1981, 6.

2. In the same way that "virtual reality" refers to computer-generated simulations of environments in the "real world" (or in imaginary worlds), "virtual sonic space" refers, in the present context, to sound-generated simulations. A sonically virtual space, then, is both absent and present at the same time—that is, its presentness derives from our imaginations, when we hear sounds that we interpret as signs of an actual environment.

3. The reason that reverb, in unmusical settings, is more common than echo is that the acoustic production of echo—that is, *distinct* sound reflections—requires a very specific architectural design to avoid the multitude of sound reflections (which characterize reverb) that are usually created from the various obstacles in a given environment.

Notes

4. The sound of the reverb is determined by several other factors as well, such as whether there are any obstacles within the room and what shape and textures these obstacles have.

5. Moore introduced his "sound box" model in *Rock: The Primary Text* (1992, revised 2001).

6. Simultaneously, but independent of Moore, Anne Danielsen (1993, 1997) presented a similar three-dimensional model, which she labeled with the Norwegian word *lydrom*, meaning sound room, to evoke an actual or concrete enclosed environment, such as the space of the recording studio. Danielsen then contrasted her concrete three-dimensional model with the excesses of the virtual environments embedded in the productions of Prince's *Diamonds and Pearls* album (1991). This contrast—comparing the qualities of an actual space against the qualities of an abstract space—is a relevant and useful distinction in the context of the present discussion.

7. For a thorough discussion of how room ambience has been exploited to aesthetic ends in the process of making records, see Horning 2012.

8. For an introduction to plate and spring reverb, see White 2003, 195–196.

9. According to Albin J. Zak III, Les Paul was one of the first to experiment with moving the playback head to achieve different delay times (Zak 2010, 317).

10. For a discussion of this process, see, e.g., Pohlmann 2000, 593; Proakis and Manolakis 1996, 1–6.

11. Albin J. Zak III helpfully identifies several other examples of how musical spatiality is experimented with in predigital recordings; see, e.g., Zak 2001, 79–83; 2012.

12. In the digital era, most professional studios abandoned plate reverb in favor of the new technologies (White 2003, 196), and Quantec's model QRS from 1982 was among the first and most popular of the digital reverb effects. Others were EMT's model 250, which was introduced in 1976; Lexicon's model 224 from 1978; and AMX's model RMX16 from 1981.

13. A case in point is Audio Ease's Altiverb 7 plug-in, which supplies reverb via samples of a variety of actual spatial environments; these reverb presets are then represented in the Altiverb browser by photographs of these environments. One can thus choose among the different reverbs by selecting pictures of various concert halls, cathedrals, recording studios, stadiums, clubs, domestic spaces, cars, and outdoor spaces. Moreover, one can place one's sounds wherever one wants within these spaces. For a demonstration of how this reverb plug-in works, see http://www.audioease.com/Pages/Altiverb/.

14. If one walks through the corridor in a crowded hotel, or if one walks along the street in the city at a Saturday night, one might hear different sounds from different

rooms or buildings at the same time, but one is never *within* these spaces simultaneously, which describes the effect of "Get Out of My House."

15. Smalley (2007) first introduced Hall's classification of distances to music analysis, and we found the approach particularly useful to our own work with musical spatiality as well (see Smalley 2007, 41).

16. For additional description of the differences between analog and digital delay, see Brice 2001, 117–120. Other analog forms of delay, such as the Bucket Brigade delay (BBD), were solid state, but common to all analog delays was the deterioration of the repeated sound's quality.

17. For a description of how Hugh Padgham created the gated reverb effect using analog technology, see Cunningham 1998, 322–328; Zak 2001; 79–81, or hear Padgham explain it himself in a TV interview with George Shilling, available at http://www.recordproduction.com/hugh_padgham.htm (accessed June 8, 2015).

18. For a description of how digital gated reverb functions, see White 2003, 202–204.

19. For a description of reverse reverb, see White 2003, 202–204.

20. See also Moylan 2012.

21. For instance, Portishead's "Half Day Closing" (*Portishead*, GO! Beat Records) from 1997 is an example of sonically simultaneous virtual spaces, which were produced by recording natural environments and using analog processing effects. For an analysis of this track, see Brøvig-Andersen 2007; Brøvig-Hanssen and Danielsen 2013.

22. Konnakol is the basic language related to the sounds of the Indian *mridangam* drum, in which each syllable (*solkattu*) represents different drum strokes, but Kannakol has also become an individual art form. See Young 1998, 12, 23.

23. See, e.g., Hawkins 2002; Danielsen 1997.

24. See, e.g., Hamlow 2003; Mitchner 2011.

3 The Instrument Formerly Known as the Machine: Hyperaccuracy and Sonic Richness in Prince's "Kiss"

1. The software Csound is a derivative of Mathews's initial program MUSIC, which is still in use today (Manning 2004, 357–358).

2. Unit generators were added to MUSIC III in 1960 (Holmes 2012, 273).

3. For a discussion of some of the different meanings of the term "sampling," see Kvifte 2007, 106–108.

4. For a short history of the Mellotron, see Reid 2002.

5. Digital *sampling* synthesis is different from the classic fixed-waveform synthesis, or digital *sound* synthesis, in that, instead of scanning a small, fixed wavetable containing one cycle of a waveform, it scans a large wavetable that contains the thousands of individual cycles of a prerecorded sound (Roads 1996, 117).

6. The first commercial drum machine, the Sideman, was produced by Wurlitzer in 1959; it was based on electromechanical technology and was to be installed in their organs. The Sideman was tube-based (see http://www.synthmuseum.com/wurl/wursideman01.html, "Sideman" [accessed September 14, 2014]). Later rhythm machines using transistor technology were installed on electric organs to supply accompaniment to the organist. Ace Tone released the first drum machine with presets in 1967. The founder of Ace Tone later established the company Roland, which delivered a series of classic drum machines with unique, artificially sounding "drums" that made a profound sonic mark on electronic dance music in the predigital era (Reid 2004).

7. Because of memory limitations, the LM-1 did not offer cymbal sounds. Cymbals were thus usually overdubbed live.

8. Oberheim introduced the DMX in 1980 (Delton 2012a), which also featured digitally sampled sounds and a "swing" feature similar to the one found on the Linn machines. It became very popular in its own right, especially in the growing hip-hop scene. Other early digital drum machines included Sequential Circuit's DrumTraks and Tom, the E-mu Drumulator, and the Yamaha RX11.

9. See MIDI Manufacturers Association 2014, on which we relied for the following summary.

10. The voltage-controlled circuits of analog sequencers are temporally unstable at the micro level, so even if all of the events were located on the metric grid, the millisecond precision that digital technology provides was not possible with analog sequencers.

11. For details on the LinnDrum 9000, see Vintage Synth Explorer 2014. The LinnDrum 9000 featured a sampling opportunity as well, which means that the sequencer could use sounds other than the eighteen prefabricated choices in its sound bank.

12. According to an earlier interview in *Mix Magazine* with David Z, the delay unit was set to 150 milliseconds (Daley 2001). Interestingly, in the final version the delay time is shorter, which might indicate that, at some point in the process, the tempo of the song was increased.

13. According to the interview in *Mix Magazine* (Daley 2001), David Z used a Kepex noise gate to achieve this effect, with a technique commonly referred to as sidechain gating. Also see the comments following the article Whitwell 2004.

Side-chaining is today often used to control compressors in hip hop and electronic dance music (EDM). In this case, the compressor uses a signal other than the main input—for example, the bass drum or the lead vocal—to control the amount of compression. The result is a characteristic "ducking" effect: the level of one audio signal (the main output) is reduced by the presence of another audio signal (the side-chain).

14. For a discussion of the use of surreal spaces in popular music, see Brøvig-Hanssen and Danielsen 2013.

15. According to Zeiner-Henriksen (2010a, 129–139), electronic dance music artists often strove for this effect; in particular, the sound of the bass drum of the Roland TR-909 used to be manipulated so that it had a descending pitch movement that added a particular feel to the groove.

16. According to David Z (Buskin 2013), there were nine tracks, which implies that the guitar hook and the wah-wah guitar were on different tracks, or that the backing vocals were placed on two tracks and panned to either side. For the purposes of this analysis, we will divide the sounds into ten tracks (two tracks for the backing vocals and two for the guitar sounds).

17. In "Kiss," according to David Z, "There was no reverb on anything else, just the kick" (Buskin 2013). It is unclear whether "anything else" refers to the groove alone or the production as a whole.

18. This was a central aspect of the negative feedback Prince received about "Kiss" from his record label: "The A&R guy said it sounded like a demo. ... no bass, no reverb" (David Z, quoted in Daley 2001); "[Prince] basically forced Warner to put it out" (David Z, quoted in Buskin 2013).

19. It is not clear whether the backing vocals had been processed as part of the Mazarati recording session—that is, before the mixing process. As we have suggested elsewhere, because of the combination of their voluminous, high-intensity sound and lack of spatial reverberation, one might speculate that they were processed with gated reverb. For further discussion of the possibly surrealistic spatiality of this song, see Brøvig-Hanssen and Danielsen 2013.

20. In fact, different sound shapes can be seen to provide the grooves with an aspect of timing variation, as a consequence of the variable distribution of energy within the sounds. If the energy of the sound is actually located after its onset (on the grid), the sound may be experienced as a little late, and so forth. The relationship between sound and timing at the microlevel of a groove is also discussed in Danielsen 2012.

21. Interestingly, in an article in *Sound on Sound* from October 1999, nearly fifteen years after the production of "Kiss," this absence of variation in sound is still mentioned as the main problem when one seeks realistic sequenced drum parts: "A main

problem with many sampled sound sets is that they do not reflect the ways in which the sound of real percussion instruments varies depending on the force with which they're struck" (Inglis 1999). This uniformity is particularly acute with hi-hat strokes: "Standard drum kit sets, particularly those conforming to the general MIDI drum map, suffer persistent problems. Perhaps the most obvious of these is the use of only three different hi-hat sounds—open, closed and pedal—when real drumming makes use of a continuous range of sounds from quiet to soft, from tight closed to open" (ibid.). An important aspect of the machinelike character of many sequencer-based grooves in the early 1980s, thus, was exactly this absence of variation among strokes (in addition to the general lack of sonic richness and subtlety).

4 The Rebirth of Silence in the Company of Noise: Portishead Going Retro

1. "High fidelity"—literally, a musical reproduction's fidelity or truth to a source—is in fact a problematic notion, because, from the term alone, it is not obvious what this "source" is. There are two possibilities: either the reproduced sounds are true to the original sounds, or the sonic event represented by the reproduced sounds is true to an external event. Here, we will reserve the term for the mediated sounds' fidelity, or transparency, to the original *sounds*, and the same applies to our discussion of lo-fi music.

2. Symes (2004, 73) and Morton (2006, 94) seem to disagree slightly about when the term was introduced.

3. Quoted in Thompson 1995, 144.

4. See Day 2000, 33, for a discussion of the ways in which recording engineers coped with the phonograph's inability to produce perfect fidelity to its sound sources.

5. The British label Decca Records claimed to have developed "full frequency range recording" (FFRR) already in 1945, but the audible frequency spectrum was in fact not entirely covered; the label's machines could only capture frequencies from 100 Hz to 14 kHz (Day 2000, 19).

6. This CD is discussed by Greg Milner (see Milner 2010, 197).

7. Sir Edward Montague Compton Mackenzie (1883–1972) was a famous English-born Scottish novelist.

8. For other examples of popular music featuring exaggerate vinyl noise, see Auner 2000; Clarke 2007; Link 2001.

9. Adrian Utley first became an official member of Portishead shortly after they released their debut album, *Dummy*, but even here he played a significant role in composing, producing, and playing guitar on several tracks.

10. Central to the trip-hop movement was the Wild Bunch, a Bristol DJ team formed in 1982 by Grant Marshall (Daddy G), Andrew Vowles (Mushroom), and Robert Del Naja (3D)—who later started their Massive Attack project—as well as Nellee Hooper, Miles Johnson, and Claude Williams. The record label Mo' Wax, founded by James Lavelle and Tim Goldsworthy in 1992, also contributed to fostering this musical style (for more information on the trip-hop movement, see Johnson 1996).

11. Trip-hop artist DJ Shadow, for example, said this about his debut hit "In Flux" (*Camel Bobsled Race*, Mo' Wax, 1998): "I don't take drugs ... people told me the music took you somewhere that may be similar. It's the track I'd always wanted to do" (quoted in Pemberton 1994). This is also how Tom Rowlands from the British electronic duo the Chemical Brothers (then known as the Dust Brothers—not to be confused with the Los Angeles–based Dust Brothers) interprets DJ Shadow's music: "I really like DJ Shadow. It's a really weird way of approaching hip-hop. I like records that make you feel like you're on drugs but you're really not" (ibid.).

12. *Dummy* reached a wide range of listeners in the United States as well as in Britain and the rest of Europe, and the following year they received the prestigious Mercury Music Prize for the album [see BBC News 2015]).

13. Tricia Rose makes this point in terms of hip hop as well: "The quality of sound found in these 1960s and 1970s soul and funk records are as important to hip-hop's sound as the machines that deconstruct and reformulate them" (Rose 1994, 78).

14. Joseph Schloss quotes hip-hop producer Domino, who also samples his own music to create a medium signature from the past: "I think there's a lot of people out there playing stuff that doesn't sound ... like the sounds are either—to me—too new, or just sound real generic, you know? So the stuff that I did that's live, I kinda want it to sound like it's a sample, in a sense" (Schloss 2004, 71).

15. In several interviews, Portishead describes their use of diverse techniques to make sounds appear to be old samples. See, e.g., O'Sullivan 1998, 77; Curwen 1999, 75.

16. Note that 24-bit depth represents 16,777,216 intervals of measurement of sample voltage, while 12-bit depth represents just 4,096 intervals, and 8-bit represents just 256 intervals. See chapter 1 for an explanation of how the digital conversion of sound works.

17. Portishead's juxtaposition of a previous medium's signature with a clear acknowledgment of the present medium complies with a form of remediation described by Jay David Bolter and Richard Grusin in which the digital medium exposes a previous medium in addition to itself and thus highlights the differences therein (Bolter and Grusin 2000, 44–50). Bolter and Grusin contrast this practice with another form of remediation, in which an older medium is remediated in digital form in such a way that it leaves the digital medium more or less transparent

(ibid.). This latter form of remediation complies with the nostalgic strategies in popular music that foreground both the qualities and the limitations of technology from the past while dismissing today's technology altogether—that is, using vinyl noise and old sound equipment *throughout* the production, rather than in contrasting passages, to evoke another era altogether.

18. Although "Strangers" does not have a music video, it is perfectly possible to reflect a commitment to new technology alongside a fetishization of the old in film, as Quentin Tarantino's thriller *Death Proof* (2007) demonstrates. Throughout this film, signatures of old technologies include black-and-white images, images with poor color quality and low resolution, jump cuts, and other medium "flaws" that the digital film medium has made obsolete. Yet *Death Proof* insists upon its contemporaneity regardless—for example, in one of its black-and-white shots, a woman pulls a mobile phone from her pocket. Like "Strangers," *Death Proof* reveals itself to be grounded in the present while drawing on (and subsumed in) the past.

19. See Askerøi 2013 for a thorough discussion of how musical sounds in pop productions might be used precisely because they function as "sonic markers" of already ascribed meanings—that is, they bear with them references to a particular context in which they are or were understood in a particular way.

5 Cut-Ups and Glitches: The Freeze and Flow of Los Sampler's and Squarepusher

1. Kid Simius (José Antonio García Soler), personal email to the author, August 26, 2008. Although Kid Simius was at the time unaware of it, this expression recalls Gang Starr's "Step in the Arena," where Guru raps: "If a beat was a princess, I would marry it."

2. In addition to his literary cutting and pasting (he spliced together pieces of his own writings), Burroughs also experimented with *sonic* cut-up poetry, splicing bits of tape from a magnetic tape recorder (see Burroughs 2009; Lydenberg 1994).

3. *Williams Mix* was created in 1952 as part of Cage's Project of Music for Magnetic Tape, whose aim was to explore tape as a medium for creating music in and of itself (see, e.g., Holmes 2012, 100–105).

4. Audio files require a large amount of processing power from the computer, and in the 1990s, when processing power was still quite expensive, the computer's playback of audio files often ended in hiccups or crashes due to buffer underruns.

5. *Glitch*, which originates in the Yiddish term *glitshn* ("to slide or skid"), is often used to describe a technological defect or error (often in computer software or hardware).

6. See also Marclay and Tone 2009, 341–347; Kelly 2009, 236, 254–257; Sangild 2004, 261. This technique is, for example, used on their commercially successful second album *Systemisch*, released by the German independent record label Mille Plateaux in 1994.

7. There now exist several DAW plug-ins (such as Effectrix by Sugar Bytes, Glitch by Illformed, and Livecut by MDSP & Smart Elextronix) with presets that create various glitch effects, such as "stutter," "tape stop," "warpcut," "repeat," "silence," "bitcrusher," "reverser," "stretcher," "vinyl noise," "lo-fi," and so forth.

8. Señor Coconut's breakthrough album, *El Baile Alemán* (2000), consists of "electrolatin" remixes of songs by the German techno pioneers Kraftwerk, and its last release, *Around the World with Señor Coconut and His Orchestra* (2008), electrolatinizes international pop hits such as "Around the World" by the French electronic music duo Daft Punk, "Sweet Dreams" by the British pop/rock duo Eurythmics, and "Kiss" by the US pop icon Prince.

9. "La Vida es Llena de Cables" is in the recording's liner notes translated with "Life Is Full of Cables." Another (Chilean) version of "La Vida es Llena de Cables (Son Disco Duro)" appears on Uwe Schmidt's 2008 Señor Coconut album *Around the World* (Nacional Records).

10. While *digital* signal dropouts consist of digital silence, this is not a defining feature of signal dropouts in general; one can also cut an analog tape and insert a blank sequence of tape within it, and those sequences of analog signal dropouts always leave behind some hum and crackle. Moreover, although the signal dropout of a digital track will result in digital silence if heard in isolation, this is not always the case when one is listening to the track as a whole; other instruments may still sound during the dropout.

11. *Buena Vista Social Club*, released in 1997, was produced by Cuban musician Juan de Marcos González and American guitarist Ry Cooder; it involves Cuban musicians performing traditional music in the vein of the Havana music scene of the 1940s and 1950s.

12. Notable glitch artists include Oval, Matmos, Pan Sonic, Alva Noto (Carsten Nicolai), Autopoieses, Farben, Frank Bretschneider, Kid 606, Kit Clayton, Pole, SND, Vladislav Delay, Authecre, Aphex Twin, and Squarepusher. Independent record labels that have been central to the global glitch movement include Mego (Vienna), Mille Plateaux (Frankfurt), Thrill Jockey (New York), Touch (London), and Warp (Sheffield/London). For discussions of glitch music, see, e.g., Bates 2004; Cascone 2000; Kelly 2009; Prior 2008; Sangild 2004; Young 2002. See also Harkins 2010, for a discussion of two artists—Akufen and Todd Edwards—who have fused the glitch aesthetics with house music.

13. This independent British record label, which is located in London, was founded in Sheffield in 1989.

14. This dichotomy also recalls the relationship between "textual silence" and "medium silence" discussed by Danielsen and Maasø. While textual silence is part of the written or performed work, medium silence results from an error in the medium (Danielsen and Maasø 2009, 129–132).

15. This duck–rabbit figure, which Ludwig Wittgenstein introduced in his *Philosophical Investigations* (1953), and which was later used by Ernst H. Gombrich in his *Art and Illusion: A Study in the Psychology of Pictorial Representation* (1960), alternately appears as one or the other creature—what might be interpreted as a bird's bill might also be seen as the ears of a rabbit, and what might be interpreted as the duck's occiput might also be seen as the rabbit's nose, and so on. See Hutcheon 1994, 59–61.

16. It is important to note here that an unmanipulated sound consists of an "attack" (the sound's onset), a "decay" (the transition from the attack to the sustain), a "sustain"/"steady state" (the middle section of a sound), and a "release" (the sound's offset or fadeout).

17. This rhythmic quality evokes the "electric boogie," in which dancers freeze and then flow in succession.

6 Seasick Computers: Microrhythmic Manipulation in the Era of Endless Undo

1. For analysis and discussion of Parliament's funk grooves, see Danielsen 2006, chap. 7.

2. The musicians in James Brown's band, for example, were known for their extreme temporal control, and the musicians in the band Toto are still renowned for their prodigious ability to play in time and for their outstanding microtiming abilities. Toto supplied the studio musicians for Michael Jackson's *Thriller* (Epic, 1983), for example.

3. Digitally stored audio represents a lot of data, and processing audio is demanding. Up to the late 1980s, then, consumer-level computers such as the Atari ST and Apple Macintosh had only enough computing power to handle MIDI data. By the late 1980s, however, computers could also handle small amounts of digital audio data. In 1992 Cubase Audio became the first sequencer program to offer audio support in addition to MIDI sequencing (Musicradar 2011).

4. For a detailed description of this technique, see Burgess 2014, 138.

5. For a detailed analysis of Brandy's "What about Us," see Carlsen and Witek 2010.

6. The vocal sample is from "One for the Treble" (Tuff City, 1984), performed by Davy DMX.

7. This option is proposed by Kristoffer Carlsen (2007) in his analysis of the groove.

8. See Johnson 2005 for an overview of the equipment used in Snoop Dogg's recording studio at the time.

9. Using samples as part of such a glitch aesthetic became a fad in certain avant-garde electronica circles in the late 1990s.

10. Multiple onsets of a particular beat falling within the boundaries of the perceived beat bin will be heard as merging into one beat, whereas onsets falling outside these boundaries will be heard as belonging to another category—namely, that of "not part of the beat" (Danielsen 2010b, 29–32).

11. Another example of a clash between subdivisions in programmed grooves from the same time period is "Nasty Girl" by Destiny's Child (*Survivor*, Columbia, 2002). For a thorough analysis of this groove, see Danielsen 2015.

12. For analysis of this and other structural aspects of rhythm in electronic dance music, see Butler 2006 and Zeiner-Henriksen 2010b.

13. For studies in the microtiming of played grooves in jazz and African-American popular music, see, e.g., Butterfield 2010; Danielsen 2006, 2012; Iyer 2002; Keil 1994; Prögler 1995.

14. A similar though less radical example is James Brown's "I Got the Feelin'" (King, 1968); see Danielsen 2006, 165–166.

15. A similar rhythmic feel is found in the vocal parts of the chorus of "Nasty Girl" by Destiny's Child, where the rhythmic fabric also consists of a mix of duple and triple subdivision. For detailed analysis, see Danielsen 2015.

16. Recent research within neuroscience and music psychology has shed light on the role of the sensorimotor system in the perception and learning of timing. Central here is the fact that when one reproduces microtemporal relationships, one does not recall their temporal features as such but rather the bodily feeling of doing so. This is supported by research into the tight action-perception coupling in rhythm (see, e.g., Large 2000; Chen, Penhune, and Zatorre 2008; Repp 2005; Repp and Su 2013). From this perspective, a particular timing pattern comes forward first and foremost as a particular bodily feeling. In accordance with this, one might suggest that a pattern of free rhythm like the bass riff of "Can I Get a Flicc Witchu" is both perceived and remembered as a gesture, and reproduced by way of the sensorimotor system. What is reproduced in Snoop's rap, then, is not a certain virtual organization of durations in an isochronometric system but a particular gestural pattern produced by the bass that is encoded as a feeling in the body.

7 Autotuned Voices: Alienation and "Brokenhearted Androids"

1. The expression "brokenhearted androids" is a quote from a review by Richards (2008).

2. The analog vocoder must not be confused with the phase vocoder (PV), which is based on windowed spectrum analysis and converts a sampled input signal into a time-varying spectral format (Roads 1996, 148, 566–567; 2001, 253–254).

3. A footnote in the article specifies that this song was "the first commercial recording to feature the audible side-effects of Antares' Auto-Tune software used as a deliberate creative effect," and moreover, that at the time of its publication in February 1999 "the producers ... were apparently so keen to maintain their 'trade secret' process that they were willing to attribute the effect to the (then) recently-released Digitech Talker vocoder pedal" (Sillitoe 1999).

4. For a comparison of Auto-Tune and Melodyne, see Walden 2007. Recent versions of both programs offer high-quality sound and a wide range of creative features. According to music producer Gary Bromham (personal communication), the early and more lo-fi versions of Auto-Tune had more "character" and are still used, particularly when one wants to produce special effects.

5. We want to thank Gary Bromham for making us aware of these different forms of usage.

6. According to Mike Dean, the mixer of the song "Love Lockdown" from *808s and Heartbreak*, West simply fell in love with the Auto-Tune effect; see Rogerson 2008a.

7. In the liner notes to Bon Iver's *Blood Bank* (Bon Iver 2014), the song is described as a meditative R&B a capella, despite the fact that there is no groove in the song. The only traits that even remotely evoke R&B are the digital melismas in the improvised voices toward the end of the song.

8. Following "Lost in the World," Bon Iver also contributed vocals to the tracks "Monster" and "Dark Fantasy" on Kanye West's album *My Beautiful Dark Twisted Fantasy* (Roc-A-Fella/Def Jam, 2010).

9. There is little experimental research into this topic within the psychology of language, but one study found that poems with a relatively high frequency of so-called plosive sounds (stop consonants) are more likely to be heard as expressing a pleasant and active mood, whereas a relatively high frequency of nasal sounds signals an unpleasant mood with low activation. When one applies Auto-Tune to a voice, the ratio between nasal and plosive sounds is clearly adjusted in favor of nasal sounds. The poetry study, to our knowledge, has never been repeated (Auracher et al. 2010).

10. The original quote from Butler is as follows: "gender parody reveals that the original identity after which gender fashions itself is an imitation without an origin"

(Butler 1990, 138). An essay by Alexandra Apolloni presents a more pessimistic cultural-critical reading of Lady Gaga's use of Auto-Tune on "Starstruck," which Apolloni reads as a self-conscious performance of "vocal damage," associating it with a body that has been reduced and transformed to (commercial) sound. She writes, "Lyrically, the song equates Gaga's body with the technology of dance music, of recording and mixing. 'Put your hands on my waist, pull the fader,' she sings, 'Would you make me number one on your playlist?' This conflation makes it seem as though her body has become sound, as thought [sic] it has become the ultimate, technologically-mediated body of celebrity, a body as a faltering sound object that you can play and put on your playlist" (Apolloni 2014, 203).

11. These various "depth models," as Fredric Jameson (1984) once called them, were under scrutiny during the wave of so-called postmodern philosophy, literary theory, and art theory in the later decades of the previous century. Much of the criticism of them centered around what was termed the "death" of the subject: the autonomous bourgeois subject was revealed to be a myth that was about to dissolve, to be replaced by a conception of practices, discourses, and textual play.

12. See also Danielsen 1997 for a discussion of how Prince withdraws from authentic expression on the album *Diamonds and Pearls* (Warner, 1991) through a combination of vocal mannerism, obvious technological mediation of the voice, and cultivation of different vocal personas throughout the album.

13. For a discussion of the primitivist understanding of black musical roots in the rock mainstream, see Danielsen 2006, chap. 2.

14. For a discussion of the close connection between a perceived raw, immediate (in the sense of unmediated) sound and authenticity in rock, see Simon Frith's classic essay "Art versus Technology: The Strange Case of Popular Music" (Frith 1986).

8 Popular Music in the Digital Era

1. Schafer introduced the term in *The New Soundscape: A Handbook for the Modern Music Teacher* (1969). He admitted to exploiting his new linguistic construction's associations and "intending it to be a nervous word. Related to schizophrenia, I wanted it to convey the same sense of aberration and drama" (Schafer 1977, 91). In our use of the term here, we do not mean to endorse Schafer's misgivings but instead hope to evoke its most literal meaning alone—that is, the split ("schizo-") of sounds ("-phonia") caused by the recording medium.

2. Of course, these eras were not based exclusively on these respective technologies—for example, digitally converted sounds could be stored magnetically. The terms instead point to the technological recording techniques that were new to the given era and that had a tremendous impact on the means of recording and composing music.

3. For a discussion of this particular era of schizophonia (the magnetic era), see Brøvig-Hanssen 2013b.

4. We do not mean to imply here that one inevitably compares recorded music to a live performance. The recording has been a dominant musical format for many decades. Moreover, live music today can be manipulated just as much as recorded music.

References

Altman, Rick. 1992. The material heterogeneity of recorded sound. In *Sound Theory, Sound Practice*, ed. Rick Altman, 15–31. New York: Routledge.

AMS Neve. 2000. *The RMX-16. Digital Reverberation System User Manual 527-530*, issue 3, 45. Accessed April 16, 2015. http://www.ams-neve.info/rmx16/manuals/Rmx_User_Iss3.pdf.

Anderson, Lessley. 2013. Seduced by "perfect" pitch: How auto-tune conquered pop music. *Verge* (February 27). http://www.theverge.com/2013/2/27/3964406/seduced-by-perfect-pitch-how-Auto-Tune-conquered-pop-music.

Anderton, Craig. 2006. How vocoders work. *PAiA Corporation Website*. Accessed April 16, 2015. http://www.paia.com/ProdArticles/vocodwrk.htm.

Antares Audio Technology. 2015. A brief history of Antares. Accessed April 17, 2015. http://www.antarestech.com/about/history.php.

Apolloni, Alexandra. 2014. Starstruck: On Gaga, voice, and disability. In *Lady Gaga and Popular Music: Performing Gender, Fashion, and Culture*, ed. Martin Iddon and Melanie L. Marshall, 190–208. New York: Routledge.

Askerøi, Eirik. 2013. Reading pop production: Sonic markers and musical identity. PhD diss., University of Agder.

Auner, Joseph. 2000. Making old machines speak: Images of technology in recent music. *Echo: A Music-Centered Journal* 2 (2). http://www.echo.ucla.edu/Volume2-Issue2/auner/auner.pdf.

Auracher, J., S. Albers, Y. Zhai, G. Gareeva, and T. Stavniychuk. 2010. P is for happiness, N is for sadness: Universals in sound iconicity to detect emotions in poetry. *Discourse Processes* 48 (1): 1–25.

Bates, Eliot. 2004. Glitches, bugs, and hisses: The degeneration of musical recordings and the contemporary musical work. In *Bad Music: The Music We Love to Hate*, ed. Christopher. J. Washburne and Maiken Derno, 275–293. New York: Routledge.

Baudrillard, Jean. 2004. *Simulacra and Simulation*. Ann Arbor: University of Michigan Press.

BBC News. 2015. Highs and lows of the Mercury Music Price. Accessed April 16, 2015. http://news.bbc.co.uk/2/shared/spl/hi/guides/456900/456975/html/nn1page5.stm.

Ben. 2006. Songs with fake scratches, vinyl noise etc. *Ilxor* (September 14). http://www.ilxor.com/ILX/ThreadSelectedControllerServlet?showall=true&bookmarkedmessageid=33&boardid=41&threadid=54390.

Benjamin, Walter. 1968. The work of art in the age of mechanical reproduction. In *Illuminations: Essays and Reflections*, ed. Hannah Arendt, 217–242. New York: Schocken Books. Originally published 1936.

Bergh, Arild, and Tia DeNora. 2009. From Wind-up to iPod: Techno-cultures of listening. In *The Cambridge Companion to Recorded Music*, ed. Nicholas Cook, Eric F. Clarke, Daniel Leech-Wilkinson and John Rink, 102–115. Cambridge: Cambridge University Press.

Bolter, Jay David, and Richard Grusin. 2000. *Remediation: Understanding New Media*. Cambridge, MA: MIT Press.

Bon Iver. 2014. Liner notes to *Blood Bank*. Accessed April 16, 2015. http://boniver.org/albums/.

Bregman, Albert S. 2001. *Auditory Scene Analysis: The Perceptual Organization of Sound*. Cambridge, MA: MIT Press.

Brice, Richard. 2001. *Music Engineering*. Oxford: Newnes.

Brøvig-Andersen, Ragnhild. 2007. Musikk og mediering: Teknologi relatert til sound og groove i trip-hop-musikk [Music and mediation: Technology, sound, and groove in trip-hop music]. MA thesis, University of Oslo.

Brøvig-Hanssen, Ragnhild. 2010. Opaque mediation: The cut-and-paste groove in DJ Food's "break." In *Musical Rhythm in the Age of Digital Reproduction*, ed. Anne Danielsen, 159–175. Farnham: Ashgate.

Brøvig-Hanssen, Ragnhild. 2013a. Music in bits and bits of music: Signatures of digital mediation in popular music recordings. PhD diss., University of Oslo.

Brøvig-Hanssen, Ragnhild. 2013b. The magnetic tape recorder: Recording aesthetics in the new era of schizophonia. In *Material Culture and Electronic Sound*, ed. Frode Weium and Tim Boon, 131–157. Washington, DC: Smithsonian Institution Scholarly Press/Rowman & Littlefield.

Brøvig-Hanssen, Ragnhild. Forthcoming. Listening *to* or through technology: Opaque and transparent mediation in popular music. Under review.

References

Brøvig-Hanssen, Ragnhild, and Anne Danielsen. 2013. The naturalised and the surreal: Changes in the perception of popular music sound. *Organised Sound* 18 (1): 71–80.

Brown, Jake. 2011. *Prince in the Studio (1975–1995)*, vol. 1. Phoenix, AZ: Amber.

Bryant, Tom. 2010. X Factor Auto-Tune scandal deepens amidst claims performances were altered during live finals. *Daily Mirror Online* (August 24). http://www.mirror.co.uk/3am/celebrity-news/x-factor-auto-tune-scandal-deepens-243155.

Bull, Michael. 2000. *Sounding Out the City: Personal Stereos and the Management of Everyday Life*. Oxford: Berg.

Bull, Michael. 2007. *Sound Moves: iPod Culture and Urban Experience*. London: Routledge.

Burgess, Richard J. 2014. *The History of Music Production*. New York: Oxford University Press.

Burroughs, William S. 2009. The invisible generation. In *Audio Culture: Reading in Modern Music*, ed. Christoph Cox and Daniel Warner, 334–340. New York: Continuum. Originally published 1962.

Bush, Kate. 1982. "Get Out of My House": Reprint of interview in the *Garden* 12 (October). Accessed April 16, 2015. http://gaffa.org/garden/kate14.html#house.

Buskin, Richard. 2013. Prince's "Kiss": Classic tracks. *Sound on Sound* (June). http://www.soundonsound.com/sos/jun13/articles/classic-tracks-0613.htm.

Butler, Judith. 1990. *Gender Trouble: Feminism and the Subversion of Identity*. New York: Routledge.

Butler, Mark J. 2006. *Unlocking the Groove*. Bloomington: Indiana University Press.

Butterfield, Matthew. 2010. Participatory discrepancies and the perception of beats in jazz. *Music Perception* 27 (3): 157–176.

Cage, John. 1967. *Silence: Lectures and Writings*. Middleton, CT: Wesleyan University Press.

Carlsen, Kristoffer. 2007. Hvor er eneren? Mikrorytmikk og pulsforhold i programmerte groover innenfor kontemporær afroamerikansk populærmusikk [Where is the one? Microrhythm and pulse in programmed grooves in contemporary African-American Popular Music]. MA thesis, University of Oslo. Accessed April 16, 2015. https://www.duo.uio.no/bitstream/handle/10852/27145/Hvorxerxeneren.pdf.

Carlsen, Kristoffer, and Maria A.G. Witek. 2010. Simultaneous rhythmic events with different schematic affiliations: Microtiming and dynamic attending in two contemporary R&B grooves. In *Musical Rhythm in the Age of Digital Reproduction*, ed. Anne Danielsen, 51–68. Farnham: Ashgate.

Cascone, Kim. 2000. The aesthetics of failure: "Post-digital" tendencies in contemporary computer music. *Computer Music Journal* 24 (4): 12–18.

Chanan, Michael. 2000. *Repeated Takes: A Short History of Recording and Its Effects on Music*. London: Verso.

Chen, Joyce L., Virginia B. Penhune, and Robert J. Zatorre. 2008. Listening to musical rhythms recruits motor regions of the brain. *Cerebral Cortex* 18 (12): 2844–2854.

Clarke, Eric F. 2007. The impact of recording on listening. *Twentieth-Century Music* 4 (1): 47–70.

Cook, Richard. 1982. My music sophisticated? I'd rather you said that than turdlike! *New Musical Express*, October 23.

Creative. 2015. Product history. Accessed April 17, 2015. http://www.fairlightus.com/about.html.

Cummings, Ray. 2010. Kanye West flips Bon Iver's "Woods" into modern classic of epic proportions. *Citypages* (September 29). http://blogs.citypages.com/gimmenoise/2010/09/kanye_west_flip.php.

Cunningham, Mark. 1998. *Good Vibrations: A History of Record Production*. London: Sanctuary Publishing.

Curwen, Trev. 1999. Portishead: Living it up. *Mix* (February): 70–75.

Daley, Dan. 2001. Classic tracks: Prince's "Kiss." *Mix Magazine* (June 1). http://www.mixonline.com/mag/audio_princes_kiss/.

Danielsen, Anne. 1993. "My name is Prince": En studie i Diamonds and Pearls. MA thesis, University of Oslo.

Danielsen, Anne. 1997. His name was Prince: A study of diamonds and pearls. *Popular Music* 16 (3): 275–291.

Danielsen, Anne. 2006. *Presence and Pleasure: The Funk Grooves of James Brown and Parliament*. Middletown, CT: Wesleyan University Press.

Danielsen, Anne. 2010a. Introduction. In *Musical Rhythm in the Age of Digital Reproduction*, ed. Anne Danielsen, 1–18. Farnham: Ashgate.

Danielsen, Anne. 2010b. Here, there, and everywhere: Three accounts of pulse in D'Angelo's "Left and Right." In *Musical Rhythm in the Age of Digital Reproduction*, ed. Anne Danielsen, 19–36. Farnham: Ashgate.

Danielsen, Anne. 2012. The sound of crossover: Micro-rhythm and sonic pleasure in Michael Jackson's "Don't Stop 'til You Get Enough." *Popular Music and Society* 35 (2): 151–168.

Danielsen, Anne. 2015. Metrical ambiguity or microrhythmic flexibility? Analysing groove in "Nasty Girl" by Destiny's Child. In *Song Interpretation in 21st-Century Pop Music*, ed. Ralf Appen, Andre Doehring and Allan F. Moore, 53–72. Farnham: Ashgate.

Danielsen, Anne, and Arnt Maasø. 2009. Mediating *music*: Materiality and silence in Madonna's "Don't Tell Me." *Popular Music* 28 (2): 127–142.

Davis, Erik. 2002. Songs in the key of F12. *Wired* 10 (5). http://www.wired.com/wired/archive/10.05/laptop.html.

Day, Timothy. 2000. *A Century of Recorded Music: Listening to Musical History*. New Haven: Yale University Press.

Deleuze, Gilles. 1994. *Difference and Repetition*. London: Athlone Press.

Delton, David. 2012a. Top ten classic drum machines, 2—Oberheim DMX. *Attack* (August 1). http://www.attackmagazine.com/features/top-ten-classic-drum-machines/9/.

Delton, David. 2012b. Top ten classic drum machines, 4—Linn Electronics Linn-Drum. *Attack* (August 1). http://www.attackmagazine.com/features/top-ten-classic-drum-machines/7/.

Derrida, Jacques. 1973. *Speech and Phenomena: And Other Essays on Husserl's Theory of Signs*. Evanston: Northwestern University Press.

Derrida, Jacques. 1974. *Of Grammatology*. Baltimore: The Johns Hopkins University Press.

Derrida, Jacques. 1982. *Margins of Philosophy*. Chicago: University of Chicago Press.

Dibben, Nicola. 2009. *Björk*. Bloomington: Indiana University Press.

Diliberto, John. 1985. Kate Bush: From piano to Fairlight with Britain's exotic chanteuse. *Keyboard Magazine* (July): 56–73.

Doyle, Peter. 2005. *Echo and Reverb: Fabricating Space in Popular Music Recording, 1900–1960*. Middletown, CT: Wesleyan University Press.

Durant, Alan. 1990. A new day for music? Digital technology in contemporary music-making. In *Culture, Technology, and Creativity in the Late Twentieth Century*, ed. Philip Hayward, 175–196. London: Art Council and Libbey Press.

Ebare, Sean. 2004. Digital music and subculture: Sharing files, sharing styles. *First Monday* 9 (2). http://firstmonday.org/htbin/cgiwrap/bin/ojs/index.php/fm/article/view/1122/1042.

Eisenberg, Evan. 2005. *The Recording Angel: Music, Records and Culture from Aristotle to Zappa*. New Haven: Yale University Press.

Engel, Friedrich K. 1999. The introduction of the Magnetophon. In *Magnetic Recording: The First 100 Years*, ed. Eric D. Daniel, C. Denis Mee, and Mark H. Clark, 47–71. New York: IEEE Press.

Eno, Brian. 1999. The revenge of the intuitive: Turn off the options, and turn up the intimacy. *Wired* 7 (1). http://www.wired.com/wired/archive/7.01/eno_pr.html.

Fernando, S. H., Jr. 1994. *The New Beats: Exploring the Music, Culture, and Attitudes of Hip-Hop.* New York: Anchor Books.

Fine, Thomas. 2008. The dawn of commercial digital recording. *ARSC Journal* 39 (1): 1–18.

Fish, Stanley E. 1983. Short people got no reason to live: Reading irony. *Daedalus* 112 (1): 175–191.

Frith, Simon. 1986. Art versus technology: The strange case of popular music. *Media, Culture & Society* 8:263–279.

Frith, Simon. 1996. *Performing Rites: On the Value of Popular Music.* Oxford: Oxford University Press.

Frith, Simon. 2012. The place of the producer in the discourse of rock. In *The Art of Record Production*, ed. Simon Frith and Simon Zagorski-Thomas, 207–221. Farnham: Ashgate.

Frith, Simon, and Lee Marshall, eds. 2004. *Music and Copyright.* New York: Routledge.

Gibson, David. 2005. *The Art of Mixing: A Visual Guide to Recording, Engineering, and Production.* Boston: Thomson Course Technology. Originally published 1997.

Gibson, James J. 1982. *Reasons for Realism: Selected Essays.* Hillsdale, NJ: Erlbaum.

Gibson, James J. 1986. *The Ecological Approach to Visual Perception.* Hillsdale, NJ: Erlbaum. Originally published 1979.

Gooch, Beverley R. 1999. Building on the Magnetophon. In *Magnetic Recording: The First 100 Years*, ed. Eric D. Daniel, C. Denis Mee, and Mark H. Clark, 72–91. New York: IEEE Press.

Goodwin, Andrew. 1990. Sample and hold: Pop music in the age of digital reproduction. In *On Record: Rock, Pop, and the Written Word*, ed. Simon Frith and Andrew Goodwin, 258–273. London: Routledge.

Gombrich, Ernst H. 1969. *Art and Illusion: A Study in the Psychology of Pictorial Representation.* Princeton: Princeton University Press.

Gracyk, Theodore. 1996. *Rhythm and Noise: An Aesthetics of Rock.* Durham, NC: Duke University Press.

Grint, Keith, and Steve Woolgar. 1997. *The Machine at Work: Technology, Work and Organization*. Oxford: Polity Press.

Grossberg, Lawrence. 1992. Is there a fan in the house? The affective sensibility of fandom. In *The Adoring Audience: Fan Culture and Popular Media*, ed. Lisa A. Lewis, 50–65. London: Routledge.

Guffey, Elizabeth E. 2006. *Retro: The Culture of Revival*. London: Reaction.

Halberstam, J. Jack. 2012. *Gaga Feminism: Sex, Gender, and the End of Normal*. Boston: Beacon Press.

Hall, Edward T. 1969. *The Hidden Dimension: Man's Use of Space in Public and Private*. London: The Bodley Head.

Hamlow, Daniel J. 2003. In *The Dreaming*, Kate lets the weirdness in full blast. Record review at *Amazon.com*. Accessed April 16, 2015. http://www.amazon.com/The-Dreaming-Kate-Bush/dp/B000006MS3.

Harkins, Paul. 2010. Microsampling: From Akufen's Microhouse to Todd Edwards and the Sound of UK Garage. In *Musical Rhythm in the Age of Digital Reproduction*, ed. Anne Danielsen, 177–194. Farnham: Ashgate.

Hawkins, Stan. 2002. *Settling the Pop Score: Pop Texts and Identity Politics*. Aldershot: Ashgate.

Hawkins, Stan. 2014. "I'll bring you down, down, down": Lady Gaga's performance in "Judas." In *Lady Gaga and Popular Music: Performing Gender, Fashion, and Culture*, ed. Martin Iddon and Melanie L. Marshall, 9–26. New York: Routledge.

Heidegger, Martin. 1977. *The Question Concerning Technology and Other Essays*. Harper Colophon Books. New York: Harper & Row.

Hesmondhalgh, David. 2005. *The Cultural Industries*. London: Sage.

Hofer, Mike. 2013. Pro Tools hardware history video. Accessed April 16, 2015. https://www.youtube.com/watch?v=ENeYnkp3RrY.

Holmes, Thom. 2012. *Electronic and Experimental Music: Technology, Music, and Culture*. New York: Routledge.

Horning, Susan Schmidt. 2012. The sound of space: Studio as instrument in the era of high fidelity. In *The Art of Record Production*, ed. Simon Frith and Simon Zagorski-Thomas, 29–42. Farnham: Ashgate.

Hutchby, Ian. 2001. Technologies, texts, and affordances. *Sociology* 35 (3): 441–456.

Hutcheon, Linda. 1988. *A Poetics of Postmodernism: History, Theory, Fiction*. New York: Routledge.

Hutcheon, Linda. 1989. *The Politics of Postmodernism*. New York: Routledge.

Hutcheon, Linda. 1994. *Irony's Edge: The Theory and Politics of Irony*. New York: Routledge.

Hutcheon, Linda. 2006. *A Theory of Adaptation*. New York: Routledge.

Inglis, Sam. 1999. 20 tips on creating realistic sequenced drum parts. *Sound on Sound* (October). http://www.soundonsound.com/sos/oct99/articles/20tips.htm.

Iyer, Vijay. 2002. Embodied mind, situated cognition, and expressive microtiming in African-American music. *Music Perception* 19 (3): 387–414.

James, Robin. 2008. "Robo-diva R&B": Aesthetics, politics, and black female robots in contemporary popular music. *Journal of Popular Music Studies* 20:402–423.

Jameson, Fredric. 1984. Postmodernism, or The cultural logic of late capitalism. *New Left Review* 146:53–92.

Johansson, Mats. 2010. The concept of rhythmic tolerance—Examining flexible grooves in Scandinavian folk-fiddling. In *Musical Rhythm in the Age of Digital Reproduction*, ed. Anne Danielsen, 69–84. Farnham: Ashgate.

Johnson, Derek, and Debbie Poyser. 2001. Celemony Melodyne. *Sound on Sound* (February). http://www.soundonsound.com/sos/nov01/articles/melodyne.asp.

Johnson, Heather. 2005. The cathedral. *Mix Magazine* (July 1). http://mixonline.com/mag/audio_cathedral/.

Johnson, Phil. 1996. *Straight Outa Bristol: Massive Attack, Portishead, Tricky, and the Roots of Trip-Hop*. London: Hodder & Stoughton/Sceptre.

Katz, Mark. 2004. *Capturing Sound: How Technology Has Changed Music*. Berkeley: University of California Press.

Kaufman, Gil. 2010. Simon Cowell's "X Factor" embroiled in auto-tune controversy. *MTV News* (August 24). http://www.mtv.com/news/1646358/simon-cowells-x-factor-embroiled-in-Auto-Tune-controversy/.

Keil, Charles. 1994. Participatory discrepancies and the power of music. In *Music Grooves*, ed. Charles Keil and Steven Feld, 96–108. Chicago: University of Chicago Press.

Kelly, Caleb. 2009. *Cracked Media: The Sound of Malfunction*. Cambridge, MA: MIT Press.

Krims, Adam. 2010. The changing function of music recordings and listening practices. In *Recorded Music: Performance, Culture and Technology*, ed. Amanda Bayley, 68–85. Cambridge: Cambridge University Press.

Kvifte, Tellef. 1989. *Instruments and the Electronic Age: Toward a Terminology for a Unified Description of Playing Technique*. Oslo: Solum.

Kvifte, Tellef. 2007. Digital sampling and analogue aesthetics. In *Aesthetics at Work*, ed. A. Melberg, 105–128. Oslo: Unipub.

Kvifte, Tellef. 2010. Composing a performance: The analogue experience in the age of digital (re)production. In *Musical Rhythm in the Age of Digital Reproduction*, ed. Anne Danielsen, 213–229. Farnham: Ashgate.

Lacasse, Serge. 2000. "Listen to my voice": The evocative power of vocal staging in recorded rock music and other forms of vocal expression." PhD diss., University of Liverpool. http://www.mus.ulaval.ca/lacasse/texts/THESIS.pdf.

Large, Edward W. 2000. On synchronizing movements with music. *Human Movement Science* 19:527–566.

Leete, Norm. 1999. Fairlight computer. *Sound on Sound* (April). http://www.soundonsound.com/sos/apr99/articles/fairlight.htm.

Lessig, Lawrence. 2004. *Free Culture: The Nature and Future of Creativity*. New York: Penguin.

Lessig, Lawrence. 2008. *Remix: Making Art and Commerce Thrive in the Hybrid Economy*. New York: Penguin.

Link, Stan. 2001. The work of reproduction in the mechanical aging of an art: Listening to noise. *Computer Music Journal* 25 (1): 34–47.

Lydenberg, Robin. 1994. Sound identity fading out: William Burroughs' tape experiments. In *Wireless Imagination: Sound, Radio, and the Avant-Garde*, ed. Douglas Kahn and Gregory Whitehead, 409–437. Cambridge, MA: MIT Press.

Manning, Peter. 2004. *Electronic and Computer Music*. Oxford: Oxford University Press.

Marclay, Christian, and Yasone Tone. 2009. Record, CD, analog, digital. In *Audio Culture*, ed. Christoph Cox and Daniel Warner, 341–347. New York: Continuum International Publishing Group.

Marin, Louis. 1991. Opacity and transparence in pictorial representation. In *EST II: Grunnlagsproblemer i estetisk forskning [EST II: Basic Challenges in Aesthetic Research]*, ed. Karin Gundersen and Ståle Wikshåland, 55–66. Oslo: Norges allmennvitenskapelige forskningsråd.

Marin, Louis. 2001. Mimesis and description. In *On Representation*, 252–268. Standford, CA: Stanford University Press.

McLeod, Kembrew. 2005a. Confessions of an intellectual (property): Danger Mouse, Mickey Mouse, Sonny Bono, and my long and winding path as a copyright activist-academic. *Popular Music and Society* 28 (1): 79–93.

McLeod, Kembrew. 2005b. *Freedom of Expression (R): Overzealous Copyright Bozos and Other Enemies of Creativity*. New York: Doubleday.

McLeod, Kembrew. 2007. *Freedom of Expression: Resistance and Repression in the Age of Intellectual Property*. Minneapolis: University of Minnesota Press.

McLuhan, Marshall. 2010. *Understanding Media*. London: Routledge Classics.

MIDI Manufacturers Association. 2014. Tutorial: History of MIDI. Accessed April 16, 2015. http://www.midi.org/aboutmidi/tut_history.php.

Millard, Andre. 2005. *America on Record: A History of Recorded Sound*. Cambridge: Cambridge University Press.

Miller, Jonathan. 1995. Adrian Utley: Portishead sound shaper. *Sound on Sound* (June). http://www.soundonsound.com/sos/1995_articles/jun95/portishead.html.

Milner, Greg. 2010. *Perfecting Sound Forever: The Story of Recorded Music*. London: Granta Books.

Mitchner, Stuart. 2011. Open house: Kate Bush receives a surprise visit from Emily Dickinson. Record Review in *Town Topics* 65 (25). http://www.towntopics.com/jun2211/book.php.

Moore, Allan F. 2001. *Rock: The Primary Text; Developing a Musicology of Rock*. Hants: Ashgate. Originally published 1992.

Moore, Allan F. 2002. Authenticity as authentication. *Popular Music* 21 (2): 209–223.

Morton, David L., Jr. 2006. *Sound Recording: The Life Story of a Technology*. Baltimore: The Johns Hopkins University Press.

Moylan, William. 2002. *The Art of Recording: Understanding and Crafting the Mix*. Woburn, MA: Focal Press.

Moylan, William. 2012. Considering space in recorded music. In *The Art of Record Production*, ed. Simon Frith and Simon Zagorski-Thomas, 163–188. Farnham: Ashgate.

Musicradar. 2011. A brief history of Steinberg Cubase. *Future Music* (May 24). http://www.musicradar.com/tuition/tech/a-brief-history-of-steinberg-cubase-406132/.

Nilsen, Per. 2004. *Prince: DanceMusicSexRomance: The First Decade*. London: Firefly Publishing.

O'Sullivan, Derek. 1998. Vintage port. *Future Music* 66:76–77.

Pemberton, Andy. 1994. Trip hop. *Mixmag* (June). http://www.techno.de/mixmag/interviews/TripHop.html.

Pohlmann, Ken C. 2000. *Principles of Digital Audio*. New York: McGraw-Hill.

Prior, Nick. 2008. Putting a glitch in the field: Bourdieu, actor network theory, and contemporary music. *Cultural Sociology* 2 (3): 310–319.

Prior, Nick. 2009. Software sequencers and cyborg singers: Popular music in the digital hypermodern. *New Formations* 66 (1): 81–99.

Proakis, John G., and Dimitris G. Manolakis. 1996. *Digital Signal Processing: Principles, Algorithms, and Applications*. Englewood Cliffs, NJ: Prentice-Hall.

Prögler, Joseph A. 1995. Searching for swing: Participatory discrepancies in the jazz rhythm section. *Ethnomusicology* 39 (1): 21–54.

Read, Oliver, and Walter L. Welch. 1977. *From Tin Foil to Stereo: Evolution of the Phonograph*. Indianapolis: Howard W. Sams.

Reid, Gordon. 2002. Rebirth of the cool: Mellotron. *Sound on Sound* (August). http://www.soundonsound.com/sos/Aug02/articles/mellotron.asp.

Reid, Gordon. 2004. The history of Roland, part 1: 1930–1978. *Sound on Sound* (November). http://www.soundonsound.com/sos/nov04/articles/roland.htm.

Repp, Bruno H. 2005. Sensorimotor synchronization: A review of the tapping literature. *Psychonomic Bulletin & Review* 12 (6): 969–992.

Repp, Bruno H., and Yi-Huang Su. 2013. Sensorimotor synchronization: A review of recent research (2006–2012). *Psychonomic Bulletin & Review* 20 (3): 403–452.

Richards, Chris. 2008. "808s and Heartbreak" by Kanye West. *Washington Post* (November 24). http://www.washingtonpost.com/wp-dyn/content/article/2008/11/23/AR2008112302506.html/.

Roads, Curtis. 1996. *The Computer Music Tutorial*. Cambridge, MA: MIT Press.

Roads, Curtis. 2001. *Microsound*. Cambridge, MA: MIT Press.

Rogerson, Ben. 2008a. Kanye West loves Auto-Tune, 808 sounds, and naked girls. *MusicRadar* (October 1). http://www.musicradar.com/news/tech/kanye-west-loves-Auto-Tune-808-sounds-and-naked-girls-177589.

Rogerson, Ben. 2008b. Kanye West says Auto-Tune makes him a better singer. *MusicRadar* (December 2). http://www.musicradar.com/news/tech/kanye-west-says-Auto-Tune-makes-him-a-better-singer-185278/.

Rose, Tricia. 1994. *Black Noise: Rap Music and Black Culture in Contemporary America*. Middletown, CT: Wesleyan University Press.

Rosen, Jody. 2008. Review of *808s and Heartbreak*. *Rolling Stone* (December 11). http://www.rollingstone.com/music/albumreviews/808s-heartbreak-20081211.

Rossing, Thomas D., F. Richard Moore, and Paul A. Wheeler. 2002. *The Science of Sound*. San Francisco: Addison Wesley.

Samuel, Raphael. 1994. *Theatres of Memory*. London: Verso.

Sangild, Torben. 2004. Glitch—the beauty of malfunction. In *Bad Music: The Music We Love to Hate*, ed. Christopher J. Washburne and Maiken Derno, 257–274. New York: Routledge.

Schaeffer, Pierre. 2004. Acousmatics. In *Audio Culture: Readings in Modern Music*, ed. Christoph Cox and Daniel Warner, 76–81. New York: Continuum.

Schafer, R. Murray. 1969. *The New Soundscape: A Handbook for the Modern Music Teacher*. Toronto: Berandol.

Schafer, R. Murray. 1977. *The Soundscape: Our Sonic Environment and the Tuning of the World*. Rochester, VT: Destiny Books.

Schloss, Joseph G. 2004. *Making Beats: The Art of Sample-Based Hip-Hop*. Middletown, CT: Wesleyan University Press.

Señor Coconut. Señor Coconut and his orchestra: Around the world with Señor Coconut. Accessed April 16, 2015. http://www.senor-coconut.com/index.php?article_id=3.

Shearlaw, John. 1981. The shock of the new. *Record Mirror* (July 11).

Sillitoe, Sue. 1999. Recording Cher's "Believe." *Sound on Sound* (February). http://www.soundonsound.com/sos/feb99/articles/tracks661.htm.

Skånland, Marie S. 2011. A technology of well-being: A qualitative study on the use of MP3 players as a medium for musical self care. PhD diss., Norwegian Academy of Music.

Smalley, Denis. 1997. Spectromorphology: Explaining sound-shapes. *Organised Sound* 2 (2): 107–126.

Smalley, Denis. 2007. Space-form and the acousmatic image. *Organised Sound* 12 (1): 35–58.

Sterne, Jonathan. 2003. *The Audible Past: Cultural Origins of Sound Reproduction*. Durham, NC: Duke University Press.

Symes, Colin. 2004. *Setting the Record Straight: A Material History of Classical Recording*. Middletown, CT: Wesleyan University Press.

Théberge, Paul. 1997. *Any Sound You Can Imagine: Making Music/Consuming Technology*. Hanover, NH: University Press of New England.

Théberge, Paul. 2012. The end of the world as we know it: The changing role of the studio in the age of the Internet. In *The Art of Record Production*, ed. Simon Frith and Simon Zagorski-Thomas, 77–90. Farnham: Ashgate.

Theremin. 2014. Rhythmicon (1930). Accessed April 16, 2015. http://www.theremin.info/-/viewpub/tid/14/pid/1.

Thomas, Matt. 1998. Go lo-fi! Back to the future. *Future Music* (February): 78–80.

Thomas, Matt, and Dave Robinson. 1998. Go lo-fi! Back to the future, part 2. *Future Music* (March): 80–84.

Thompson, Emily. 1995. Machines, music, and the quest for fidelity: Marketing the Edison Phonograph in America, 1877–1925. *Musical Quarterly* 79 (1): 131–171.

Thomson, Phil. 2004. Atoms and errors: Towards a history and aesthetics of microsound. *Organised Sound* 9 (2): 207–218.

Totale, Todd. 2009. Lost classics: Kate Bush—*The Dreaming*. Record Review at *Glorious Noise* (February 12). http://gloriousnoise.com/2009/kate_bush_the_dreaming.

Toynbee, Jason. 2000. *Making Popular Music: Musicians, Creativity and Institutions*. London: Arnold.

Truax, Barry. 1988. Real-time granular synthesis with a digital signal processor. *Computer Music Journal* 12 (2): 14–26.

Truax, Barry. 2001. *Acoustic Communication*. Westport, CT: Ablex.

Tyrangiel, Josh. 2005. Auto-Tune: Why pop music sounds perfect. *TIME* (February 9). http://content.time.com/time/magazine/article/0,9171,1877372,00.html.

Vaidhyanathan, Siva. 2003. *Copyrights and Copywrongs: The Rise of Intellectual Property and How It Threatens Creativity*. New York: New York University Press.

Vintage Synth Explorer. 2014. Linn Electronics Linn 9000. Accessed April 16, 2015. http://www.vintagesynth.com/linn/linn9000.php.

Walden, John. 2007. Auto-Tune vs. Melodyne: Antares Auto-Tune 5 & Celemony Melodyne plug-in pitch-correction processors. *Sound on Sound* (March). http://www.soundonsound.com/sos/mar07/articles/at5vsmelodyne.htm.

Watkinson, John R. 1999. The history of digital audio. In *Magnetic Recording: The First 100 Years*, ed. Eric D. Daniel, C. Denis Mee, and Mark H. Clark, 110–123. New York: IEEE Press.

White, Paul. 2003. *Creative Recording, Part 1: Effects and Processors*. London: Sanctuary.

Whitwell, Tom. 2004. Tuesday is Prince day: Pt. 3: How "Kiss" was made. *Music Thing* (September 28). http://musicthing.blogspot.no/2004/09/tuesday-is-prince-day-pt-3-how-kiss.html.

Williams, Juliet. 2014. "Same DNA, but born this way": Lady Gaga and the possibilities of postessentialist feminisms. *Journal of Popular Music Studies* 26 (1): 28–46.

Young, Lisa. 1998. *Konnakol*: The history and development of *solkattu*—the vocal syllables—of the *mridangam*. MA thesis, School of Music, Victorian College of the

Arts, University of Melbourne. Accessed April 16, 2015. http://www.lisayoung.com/au/research/masters.pdf.

Young, Rob. 2002. Worship the glitch: Digital music, electronic disturbance. In *Undercurrents: The Hidden Wiring of Modern Music*, ed. Rob Young, 45–55. New York: Continuum International Publishing Group.

Zak, Albin J., III. 2001. *The Poetics of Rock: Cutting Tracks, Making Records*. Berkeley: University of California Press.

Zak, Albin J., III. 2010. Painting the sonic canvas: Electronic mediation as musical style. In *Recorded Music: Performance, Culture and Technology*, ed. Amanda Bayley, 307–324. Cambridge: Cambridge University Press.

Zak, Albin J., III. 2012. No-fi: Crafting a language of recorded music in 1950s pop. In *The Art of Record Production*, ed. Simon Frith and Simon Zagorski-Thomas, 43–55. Farnham: Ashgate.

Zeiner-Henriksen, Hans T. 2010a. Moved by the groove: Bass drums sounds and body movement in electronic dance music. In *Musical Rhythm in the Age of Digital Reproduction*, ed. Anne Danielsen, 121–140. Farnham: Ashgate.

Zeiner-Henriksen, Hans T. 2010b. The "PoumTchak" pattern: Correspondences between rhythm, sound, and movement in electronic dance music. PhD diss., University of Oslo.

Selected Discography

Bon Iver. Woods, *Blood Bank*. Jagjaguwar, 2009, compact disc.

Kate Bush. Get Out of My House, *The Dreaming*. EMI America, 1982, compact disc.

Lady Gaga. Starstruck, *The Fame*. Streamline/Universal, 2008, compact disc.

Lady Gaga. Starstruck (Official Video), 2014. Accessed April 16, 2015. https://www.youtube.com/watch?v=tVRgVpqQCtc.

Los Sampler's. La Vida es Illena de Cables, *Descargas*. Rather Interesting, 2000, compact disc.

Portishead. Strangers. *Dummy*. Go! Discs, 1994, compact disc.

Prince. Kiss. *Parade*. Paisley Park/Warner, 1986, compact disc.

Snoop Dogg. *R&G (Rhythm & Gangsta): The Masterpiece*. Geffen, 2004, compact disc.

Squarepusher. My Red Hot Car, *Go Plastic*. Warp Records, 2001, compact disc.

Squarepusher. My Red Hot Car (Girl), *My Red Hot Car*. Warp Records, 2001, compact disc.

Index

Note: Page numbers followed by "f" or "t" refer to figures and tables, respectively.

Affordances, 15–16
Altiverb 7 plug-in, 155n13
AMS RMX-16, 54
Analog sound synthesis, 44, 47, 48
Analog tape delay, 13, 30
Auditory Scene Analysis (Bregman), 89
Auner, Joseph, 61, 68, 77–78, 159n8
Auto-Tune, 117–124, 126–127, 131–132, 138–139, 145
  Bon Iver and, 121–124, 126, 131, 132, 138
  controversies and disputes regarding, 117, 129, 132, 145–146
  criticism of, 117, 121, 129, 131
  Lady Gaga and, 129–131
  Melodyne and, 165n4
  overview, 118–119
  pitch correction as vocal effect, 118–121
  Vocoder and, 118–119

"Bang Out" (Snoop Dogg), 106, 110, 110f, 111, 113, 115, 138, 145
Bell Laboratories, 45
Benjamin, Walter, 147–148
Björk, 99, 126
Bolter, Jay David, 153n6, 154n11, 160n17
Bon Iver, 123
  Auto-Tune and, 121–124, 126, 131, 132, 138
  "Woods," 121–126, 122f, 123f, 131, 138–139
Bregman, Albert S., 89
Brøvig-Andersen, Ragnhild, 70, 156n21
Brøvig-Hanssen, Ragnhild, 5, 149, 153–154nn6–7, 154n13, 156n21, 158n14, 158n19, 167n3
Buena Vista Social Club, 162n11
Burgess, Richard J., 50–51, 103, 145–146, 163n4
Bush, Kate, 21, 27–29, 34
  "Get Out of My House," 22, 27–41, 134
Butler, Judith, 129, 165n10

Cage, John, 79, 81
"Can I Get a Flicc Witchu" (Snoop Dogg), 106–111, 113–115, 137–138, 144–145, 164n16
Carlsen, Kristoffer, 106, 163n5, 164n7
Cher, 119
Cher effect, 138
Collins, Phil, 132
  "In the Air Tonight," 32
Compact disc (CD), 65–67, 82
Compositional palette, a new, 139–140

Computer Musical Instrument (CMI), 11, 47, 49
Concrete sound objects, 47
Cuban music, 84, 85, 88
Cunningham, Mark, 32
Cut-and-paste techniques, 81–99
  in analog era, 108, 133, 136–137, 139, 141, 145, 148
  in "La Vida es Llena de Cables," 83–89, 98
  in "My Red Hot Car," 90–96, 98

Danielsen, Anne, 95, 101, 155n6, 163n14, 166n12
David Z, 52–58, 135, 158nn16–18
DAWs (digital audio workstations), 1, 9, 12–13, 51, 104, 111–115, 154n12
Death Proof (film), 161n18
Decay time. *See* Reverb
Delay. *See also* Digital delay; Echo
  defined, 21, 23
  reverb and, 21, 25, 27, 29, 39–41, 54, 134, 146
Delay units, 47, 52–54, 157n12
Deleuze, Gilles, 112
Denaturalization, 129
Derrida, Jacques, 129–130
Descargas (Los Sampler's), 83, 84, 88. *See also* "La Vida es Llena de Cables (Son Disco Duro)"
Destroyed Music (Knížák), 90
Dibben, Nicola, 99, 126
Digital audio workstations. *See* DAWs
Digital delay, 21–23, 25, 27, 29–31, 40–41, 47. *See also* Delay
  vs. analog delay, 134, 156n16
  in "Get Out of My House," 29–31, 34, 35, 39–41, 134
Digital dropouts. *See* Dropouts, signal
Digital signatures, 2, 133, 139, 149. *See also specific topics*
  defined, 2
  nature of, 2, 6

opaque mediation and, 5–8
Digital sound synthesis, 44–46, 51
Digital-to-analog (DA) converter, 45
"Don't Tell Me" (Madonna), 95
Double meaning of aestheticized malfunctions, 89–97, 136–137
Doyle, Peter, 40
Dreaming, The (Bush), 28. *See also* "Get Out of My House"
Dropouts, signal, 95–98
  defined, 86
  digital vs. analog, 162n10
  in "La Vida es Llena de Cables," 86–89, 96, 98
  in "My Red Hot Car," 91–96, 98
  nature of, 86, 96–98
  in "Strangers," 78
Drum loops, 71, 73
Dummy (Portishead), 71, 72, 160n12. *See also* "Strangers"

Echo, 21, 23, 31, 40. *See also* Delay
  defined, 21, 23, 154n3
  vs. reverb, 23, 26, 154n3
  slapback, 35, 36
Echo chambers, 24
Edison, Thomas, 62, 65, 67
Edison Diamond Disc, 62
Electrolatin music, 84, 162n8. *See also* Latin American music
Electronic dance music (EDM), 14, 58, 102, 112, 113
Electronic pop stars, 90
Emulator (E-MU), 47
Eno, Brian, 78
Evans, Daniel, 117

Fairlight Computer Musical Instrument (CMI), 11, 47, 49
Fixed-waveform synthesis, 157n5
Frequency modulation (FM) algorithm, 46

Index

Frequency modulation (FM) synthesis, 11, 45–46
Frith, Simon, 127, 150

Gated reverb, 31–32, 63, 156nn17–19
Gender parody, 129
Gestural language, 109, 114–115
"Get Out of My House" (Bush), 22, 27–41, 134. *See also Dreaming*
Gibson, James J., 15–16, 148
Glitches, 81–91, 94–99, 126, 133, 136–137, 139, 145, 147, 149
  digital, 90, 97, 133, 139
  double meaning and double function (*see* Double meaning of aestheticized malfunctions)
  natural, 83, 95–97
Glitch music, 81–99, 162n12
  ways of manufacturing, 82–83
Glitch plug-ins, 162n7
Goodwin, Andrew, 68
Grint, Keith, 15
Groove
  computerizing the, 57–59
  digital, 52–57
  inner dynamics of a, 111–114
Groove-based music, 57, 102, 105, 112, 114, 144
Grooves, computer-based, 46
Grusin, Richard, 153n6, 154n11, 160n17
Guffey, Elizabeth, 76, 78

Halberstam, J. Jack, 129
"Half Day Closing" (Portishead), 156n21
Hall, Edward, 29–30
Hawkins, Stan, 127
High fidelity (hi-fi), 159n1
  digital silence and the holy grail of, 62–66
Hi-hat, 52–53, 59, 106, 110, 135, 159
Hildebrand, Andy, 118

Hutchby, Ian, 15, 16
Hutcheon, Linda, 89, 96, 147
Hypernature and hypernatural, 121–131

Inglis, Sam, 158n21
"In the Air Tonight" (Collins), 32

James, Robin, 121
Jameson, Fredric, 75

Katz, Mark, 8
Kelly, Caleb, 82, 95
"Kiss" (Prince), 51–59, 54f, 55f, 56t, 158nn17–18
Knížák, Milan, 90
Kvifte, Tellef, 143, 154n14, 156n3

Lacasse, Serge, 34
Lady Gaga, 127, 131, 148
  "Starstruck," 126–131, 128f, 139
Latin American music, 83–85, 162n8
"La Vida es Llena de Cables (Son Disco Duro)" (Los Sampler's), 83–89, 95, 96, 98, 136, 137, 145, 148
Link, Stan, 69, 79
Linn 9000, 49, 51–53
LinnDrum 9000, 157n11
LM-1, 49
Loops, 14, 73, 108
  drum, 71, 73
  tape, 25, 47–48, 105
Los Sampler's, 2, 18, 81, 83–87, 89, 133, 136, 145, 148, 149. *See also* "La Vida es Llena de Cables (Son Disco Duro)"
Low fidelity (lo-fi), 17, 61, 69–71, 78, 79
Low fidelity (lo-fi) movement, 62, 68, 136
Low-frequency oscillator (LFO), 44

Maasø, Arnt, 95, 163n14
Machine-aided vs. machine-generated, 115
Madonna, 95

Magnetic tape, 81, 103
Magnetic tape recorders, 9, 25, 64, 140–141
Marclay, Christian, 90
Marin, Louis, 77, 153n6
McLuhan, Marshall, 15
Mediation. *See also* Opaque mediation; Transparent mediation
 significance to popular music, 149–151
Medium signatures revisited, past, 66–69
Medium silence vs. textual silence, 163n14
Melismas, 123, 123f, 126
Melodyne, 120, 165n4
Microphones, 24, 32
Microrhythms, 53, 101–103, 105–115, 142, 144, 149
Microrhythmic manipulation, 104, 111, 112, 131, 133, 138
 in Snoop Dogg's music, 105–111
MIDI (Musical Instrument Digital Interface), 11, 44
 historical perspective on, 11–12, 44, 49, 50
 "Kiss" and, 57–59
 Linn 9000 and, 49, 51, 52
 quantization and, 51
MIDI code and standardizing communication, 49–51
MIDI protocol, 44, 46, 51, 58, 135
MIDI sequencer, 51, 52, 103, 104
Millard, Andre, 12, 66
Moore, Allan F., 23–24
Moylan, William, 33
Musical gestures, 95, 99, 145
Musical space, production alternatives for creating, 26–27
Musical spatiality, 21–27, 29, 34–36, 40, 41
 defined, 21
Music within the music, 83–89

"My Red Hot Car" (Squarepusher), 83, 90–96, 98, 136–137, 145, 148
"My Red Hot Car (Girl)" (Squarepusher), 90–91

Naturalist vs. interventionist work, 7–8
Naturalization, 7, 131, 133, 147–149. *See also* Denaturalization
 vs. surreality, 40–41
Noise, 43, 61–62, 67, 75, 135
 background, 56, 61, 63, 64, 67
 digital void of sound and, 43, 78
 high fidelity and, 63–65
 mingling of digital silence and analog, 75–79
 retro revivalism and, 76–79
 vinyl, 68, 69, 71, 75, 78
Noise gate, 31–32, 52

Opaque mediation, 41, 136
 digital signatures and, 5–8
 and the significance of mediation to popular music, 149–151
 vs. transparent mediation, 5–8, 68 (*see also* Transparent mediation)

Padgham, Hugh, 32, 156n17
Phonograph effects, 8, 9. *See also* Vinyl noise
Phonographs, 24, 63–64, 67, 140, 141
Pitch correction (as vocal effect). *See* Auto-Tune
Plate reverb, 13, 25, 26, 134, 155n12
 principle behind, 25
Portishead, 68, 69, 133, 160n17
 Dummy, 71, 72, 160n12
 "Half Day Closing," 156n21
 juxtaposition of old and new, 70–79
 "Strangers," 71–75, 78, 136, 145, 148, 149
 "Undenied," 75, 77–78
Portishead (Portishead), 75, 76
Postmodern art, 89, 166n11

Index

Predecay time, 23, 25, 26, 37
Primitive integration, 89
Prince, 49, 135
  "Kiss," 51–59, 54f, 55f, 56t, 158nn17–18
Prior, Nick, 14–15, 127, 143
Pulse-code modulation (PCM), 9

Quantec QRS room simulator, 29
Quantization, 44, 51

Reflections, sound, 23–26, 37, 38, 154n3. *See also* Delay; Echo; Reverb
Remediation, 154n11
Retro, defined, 76
Retrochic, 77
Retro revivalism, 76–79. *See also* Portishead
Reverb (reverberation), 23–27, 29, 33–40, 54, 56, 155nn12–13
  analog vs. digital, 25, 32, 134
  artificial, 25
  delay and, 21, 25, 27, 29, 39–41, 54, 134, 146
  vs. digital reverb, 26f, 155n12
  vs. echo, 23, 26, 154n3
  "echo chambers" and, 24
  gated, 31–32, 63, 156nn17–19
  in "Get Out of My House," 29–41, 134
  historical perspective on, 24–27
  in "My Red Hot Car," 92–94
  natural, 25–26
  overview and nature of, 23, 25–26
  plate, 13, 25, 26, 134, 155n12
  reverse, 32, 37, 41, 54, 55f, 134
  spring, 25
Reverse reverb, 37, 41, 54, 55f, 134
  defined, 32
R&G (Rhythm & Gangsta): The Masterpiece (Snoop Dogg), 106, 108, 111, 114, 115, 137–138, 144–145, 148, 149

Rhythm & Gangsta. *See* R&G (Rhythm & Gangsta)
Rhythmicon, 48
Rhythmic tolerance, 111
Roads, Curtis, 45, 47
Room simulator, Quantec QRS, 29
Rose, Tricia, 160n13

Sampler instruments, digital, 9–11, 14, 43, 44, 47–48, 52, 73
Sampling and sample-based instruments, 46–49
Sampling synthesis, digital, 44, 48, 51
Samuel, Raphael, 76, 77
Schaeffer, Pierre, 47
Schafer, R. Murray, 140, 148, 166n1
Schema-based integration, 89
Schizophonia, 62, 97, 160n1
  a new era of, 140–143
Schloss, Joseph, 160n14
Schmidt, Uwe, 83–84
Seasick rhythms, 3, 101, 105–108, 110, 113, 115, 137, 144–145
Selective deafness, 68
Sensorimotor system, 164n16
Sequencer programs, computer-based, 11–14, 71
Shining, The (King), 34–35
Sideman, 157n6
Signal dropouts. *See* Dropouts, signal
Silence, digital, 77, 145, 162n10. *See also* Dropouts, signal
  and the holy grail of "high fidelity," 62–66
  mingling of analog noise and, 75–79
  in *Portishead*, 75, 76
  silence in digital vs. analog technology, 61–62
  in "Strangers," 72, 74, 74f, 136, 149
  textual silence vs. medium silence, 163n14
  in "Undenied," 77–78

Simius, Kid (José Antonio García Soler), 81, 161n1
Slapback echo/slapback delay, 35, 36
Smalley, Denis, 7–8, 22, 156n15
Smith, Oberlin, 64
Snoop Dogg, 133
  "Bang Out," 106, 110, 110f, 111, 113, 115, 138, 145
  "Can I Get a Flicc Witchu," 106–111, 113–115, 137–138, 164n16
  microrhythms, 105–111
  R&G (Rhythm & Gangsta): The Masterpiece, 106, 108, 111, 114, 115, 137–138, 144–145, 148, 149
Son Cubano, 85, 88
Sound box, defined, 29
Sound box model, 23–24
Sound synthesis, 44–45
Source bonding, 22
Spatiality in music. *See* Musical spatiality
Spatial simultaneity, 33
Spring reverb, 25
Squarepusher, 90, 149. *See also* "My Red Hot Car"
"Starstruck" (Lady Gaga), 126–131, 128f, 139
Sterne, Jonathan, 67, 153n4
Stockham, Thomas Greenway, 66
Stone, Faith, 67
"Strangers" (Portishead), 71–75, 78, 136, 145, 148, 149
Surreal virtual spaces, 27–34, 40–41

Tape recorders. *See* Magnetic tape recorders
Tarantino, Quentin, 161n18
Textual silence vs. medium silence, 163n14
Théberge, Paul, 1, 153n1, 153n4
Thomson, Phil, 97–99
Thoreau, Henry David, 125–126

Transparent mediation, 5–8, 68, 89, 150, 151
Trip hop, 62, 70, 160nn10–11
Truax, Barry, 94
Tuning of the ear, 150
  historical listening constraints and the, 147–149

"Undenied" (Portishead), 75, 77–78
Utley, Adrian, 70, 71, 73

Vinyl noise, 68, 69, 71, 75, 78
Virtual spatial environments, 21, 23, 24, 26–27, 29, 31, 33, 35, 39–41, 154n2

Waveform, 9, 44–45, 85, 157n5
  examples of, 54f, 85f–87f, 92f–94f, 107f, 109f, 110f, 122f, 123f, 128f
Weather Report, 71, 72
West, Kanye, 120, 124
Williams, Juliet, 129
Williams Mix (Cage), 81
"Woods" (Bon Iver), 121–126, 122f, 123f, 131, 138–139
Woolgar, Steve, 15

Yamaha DX7, 45, 46
Young, Neil, 67

Z, David, 52–58, 135, 158nn16–18
Zak, Albin J., III, 31, 66
Zeiner-Henriksen, Hans T., 58, 113, 158n15
Zoned spaces, 33

www.ingramcontent.com/pod-product-compliance
Lightning Source LLC
Chambersburg PA
CBHW030654230426
43665CB00011B/1090